Cram101 Textbook Outlines to accompany:

Volds Theoretical Criminology

Thomas J. Bernard, 6th Edition

A Content Technologies Inc. publication (c) 2012.

Learning System

Cram101 Textbook Outlines is a learning system. The notes in this book are the highlights of your textbook, you will never have to highlight a book again.

How to use this book. Take this book to class, it is your notebook for the lecture. The notes and highlights on the left hand side of the pages follow the outline and order of the textbook. All you have to do is follow along while your instructor presents the lecture. Circle the items emphasized in class and add other important information on the right side. With Cram101 Textbook Outlines you'll spend less time writing and more time listening. Learning becomes more efficient.

Cram101.com Online

Increase your studying efficiency by using Cram101.com's practice tests and online reference material. It is the perfect complement to Cram101 Textbook Outlines. Use self-teaching matching tests or simulate in-class testing with comprehensive multiple choice tests, or simply use Cram's true and false tests for quick review. Cram101.com even allows you to enter your in-class notes for an integrated studying format combining the textbook notes with your class notes.

Visit **www.Cram101.com**, click Sign Up at the top of the screen, and enter **DK73DW13695** in the promo code box on the registration screen. Your access to www.Cram101.com is discounted by 50% because you have purchased this book. Sign up and stop highlighting textbooks forever.

Volds Theoretical Criminology
Thomas J. Bernard, 6th

CONTENTS

Chapter 1. THEORY AND CRIME

Crime	Crime is the breach of rules or laws for which some governing authority (via mechanisms such as legal systems) can ultimately prescribe a conviction. Individual human societies may each define crime and crimes differently. While every crime violates the law, not every violation of the law counts as a crime; for example: breaches of contract and of other civil law may rank as "offences" or as "infractions".
Theory	The word theory, when used by scientists, refers to an explanation of reality that has been thoroughly tested so that most scientists agree on it. It can be changed if new information is found. Theory is different from a working hypothesis, which is a theory that hasn't been fully tested; that is, a hypothesis is an unproven theory. The word theory also distinguishes ideas from practice.
Association	In quantitative research, the term "association" is often used to emphasize that a relationship being discussed is not necessarily causal .
Differential	In calculus, a differential is traditionally an infinitesimally small change in a variable. For example, if x is a variable, then a change in the value of x is often denoted Δx (or δx when this change is considered to be small). The differential dx represents such a change, but is infinitely small.
Differential association	In criminology, Differential Association is a theory developed by Edwin Sutherland proposing that through interaction with others, individuals learn the values, attitudes, techniques, and motives for criminal behavior. The Differential Association Theory is the most talked about of the Interactionist theory of deviance. This theory focuses on how individuals learn how to become criminals, but does not concern itself with why they become criminals.
Social learning	Social learning is learning that takes place at a wider scale than individual or group learning, up to a societal scale, through social interaction between peers. It may or may not lead to a change in attitudes and behaviour. More specifically, to be considered social learning, a process must: (1) demonstrate that a change in understanding has taken place in the individuals involved; (2) demonstrate that this change goes beyond the individual and becomes situated within wider social units or communities of practice; and (3) occur through social interactions and processes between actors within a social network (Reed et al., 2010).

Chapter 1. THEORY AND CRIME

Social learning theory	Social learning theory is the theory that people learn new behavior through observational learning of the social factors in their environment. If people observe positive, desired outcomes in the observed behavior, then they are more likely to model, imitate, and adopt the behavior themselves. Modern theory is closely associated with Julian Rotter and Albert Bandura.
Probability	Probability is a way of expressing knowledge or belief that an event will occur or has occurred. The concept has been given an exact mathematical meaning in probability theory, which is used extensively in such areas of study as mathematics, statistics, finance, gambling, science, Artificial intelligence/Machine learning and philosophy to draw conclusions about the likelihood of potential events and the underlying mechanics of complex systems. Interpretations The word probability does not have a consistent direct definition.
Deterrence	Deterrence is a theory from behavioral psychology about preventing or controlling actions or behavior through fear of punishment or retribution. This theory of criminology is shaping the criminal justice system of the United States and various other countries.
Power	The Power of a statistical test is the probability that the test will reject a false null hypothesis (i.e. that it will not make a Type II error). As Power increases, the chances of a Type II error decrease. The probability of a Type II error is referred to as the false negative rate (β). Therefore Power is equal to $1 - \beta$.
Poverty	Poverty is the lack of basic human needs, such as clean water, nutrition, health care, education, clothing and shelter, because of the inability to afford them. This is also referred to as absolute poverty or destitution. Relative poverty is the condition of having fewer resources or less income than others within a society or country, or compared to worldwide averages.
Bias	In statistics, Bias is systematic favoritism that is present in the data collection process resulting in misleading results. There are several types of statistical Bias:

Chapter 1. THEORY AND CRIME

· Selection Bias, where there is an error in choosing the individuals or groups to take part in a scientific study. It includes sampling Bias, in which some members of the population are more likely to be included than others. Spectrum Bias consists of evaluating the ability of a diagnostic test in a Biased group of patients, which leads to an overestimate of the sensitivity or specificity of the test. Funding Bias may lead to selection of outcomes, test samples, or test procedures that favor a study's financial sponsor.

· The Bias of an estimator is the difference between an estimator's expectation and the true value of the parameter being estimated. Omitted-variable Bias is the Bias that appears in estimates of parameters in a regression analysis when the assumed specification is incorrect, in that it omits an independent variable that should be in the model.

· In statistical hypothesis testing, a test is said to be unBiased when the probability of rejecting the null hypothesis exceeds the significance level when the alternative is true and is less than or equal to the significance level when the null hypothesis is true.

· Systematic Bias or systemic Bias are external influences that may affect the accuracy of statistical measurements.

· Data-snooping Bias comes from the misuse of data mining techniques.

Cultural bias	Cultural bias is the phenomenon of interpreting and judging phenomena by standards inherent to one's own culture. The phenomenon is sometimes considered a problem central to social and human sciences, such as economics, psychology, anthropology, and sociology. Some practitioners of the aforementioned fields have attempted to develop methods and theories to compensate for or eliminate cultural bias.
Capital punishment	Capital punishment is the killing of a person by judicial process as a punishment for an offense. Crimes that can result in a death penalty are known as capital crimes or capital offences. The term capital originates from Latin capitalis, literally "regarding the head" .

Chapter 1. THEORY AND CRIME

Self-image	A person's self-image is the mental picture, generally of a kind that is quite resistant to change, that depicts not only details that are potentially available to objective investigation by others (height, weight, hair color, gender, I.Q. score, etc)., but also items that have been learned by that person about himself or herself, either from personal experiences or by internalizing the judgments of others. A simple definition of a person's self-image is their answer to this question - "What do you believe people think about you?" A more technical term for self-image that is commonly used by social and cognitive psychologists is self-schema. Like any schema, self-schemas store information and influence the way we think and remember.
Wealth	Wealth is the abundance of valuable resources or material possessions, or the control of such assets. The word wealth is derived from the old English wela, which is from an Indo-European word stem. An individual, community, region or country that possesses an abundance of such possessions or resources is known as wealthy.
Risk factor	In Information security, Risk factor is a collectively name for circumstances affecting the likelihood or the impact of a security risk.

Chapter 2. CLASSICAL CRIMINOLOGY

Chicago School	In sociology and later criminology, the Chicago School was the first major body of works emerging during the 1920s and 1930s specialising in urban sociology, and the research into the urban environment by combining theory and ethnographic fieldwork in Chicago, now applied elsewhere. While involving scholars at several Chicago area universities, the term is often used interchangeably to refer to the University of Chicago's sociology department--one of the oldest and one of the most prestigious. Following World War II, a "Second Chicago School" arose whose members used symbolic interactionism combined with methods of field research, to create a new body of works.
Human Ecology	Human ecology is the interdisciplinary or transdisciplinary study of the relationship between humans and their natural, social, and built environments.
Association	In quantitative research, the term "association" is often used to emphasize that a relationship being discussed is not necessarily causal .
Crime	Crime is the breach of rules or laws for which some governing authority (via mechanisms such as legal systems) can ultimately prescribe a conviction. Individual human societies may each define crime and crimes differently. While every crime violates the law, not every violation of the law counts as a crime; for example: breaches of contract and of other civil law may rank as "offences" or as "infractions".
Differential	In calculus, a differential is traditionally an infinitesimally small change in a variable. For example, if x is a variable, then a change in the value of x is often denoted Δx (or δx when this change is considered to be small). The differential dx represents such a change, but is infinitely small.
Differential association	In criminology, Differential Association is a theory developed by Edwin Sutherland proposing that through interaction with others, individuals learn the values, attitudes, techniques, and motives for criminal behavior. The Differential Association Theory is the most talked about of the Interactionist theory of deviance. This theory focuses on how individuals learn how to become criminals, but does not concern itself with why they become criminals.
Theory	The word theory, when used by scientists, refers to an explanation of reality that has been thoroughly tested so that most scientists agree on it. It can be changed if new information is found. Theory is different from a working hypothesis, which is a theory that hasn't been fully tested; that is, a hypothesis is an unproven theory.

Chapter 2. CLASSICAL CRIMINOLOGY

The word theory also distinguishes ideas from practice.

Anomie	Anomie is a sociological term meaning "personal feeling of a lack of social norms; normlessness". It describes the breakdown of social norms and values. It was popularized by French sociologist Émile Durkheim in his influential book Suicide (1897).
Social change	Social change refers to an alteration in the social structure of a social group or society; a change in the nature, social institutions, social behaviors or social relations of a society.
	Social change is a very basic term and must be assigned further context. It may refer to the notion of social progress or sociocultural evolution; the philosophical idea that society moves forward by dialectical or evolutionary means. It may refer to a paradigmatic change in the socio-economic structure, for instance a shift away from feudalism and towards capitalism. Accordingly it may also refer to social revolution, such as the Communist revolution presented in Marxism, or to other social movements, such as Women's suffrage or the Civil rights movement. Social change may be driven by cultural, religious, economic, scientific or technological forces.
Social learning	Social learning is learning that takes place at a wider scale than individual or group learning, up to a societal scale, through social interaction between peers. It may or may not lead to a change in attitudes and behaviour. More specifically, to be considered social learning, a process must: (1) demonstrate that a change in understanding has taken place in the individuals involved; (2) demonstrate that this change goes beyond the individual and becomes situated within wider social units or communities of practice; and (3) occur through social interactions and processes between actors within a social network (Reed et al., 2010).
Social learning theory	Social learning theory is the theory that people learn new behavior through observational learning of the social factors in their environment. If people observe positive, desired outcomes in the observed behavior, then they are more likely to model, imitate, and adopt the behavior themselves. Modern theory is closely associated with Julian Rotter and Albert Bandura.
Deterrence	Deterrence is a theory from behavioral psychology about preventing or controlling actions or behavior through fear of punishment or retribution. This theory of criminology is shaping the criminal justice system of the United States and various other countries.

13

Chapter 2. CLASSICAL CRIMINOLOGY

Punishment	Punishment is the authoritative imposition of something negative or unpleasant on a person or animal in response to behavior deemed wrong by an individual or group. The authority may be either a group or a single person, and punishment may be carried out formally under a system of law or informally in other kinds of social settings such as within a family. Negative consequences that are not authorized or that are administered without a breach of rules are not considered to be punishment as defined here.
Self-image	A person's self-image is the mental picture, generally of a kind that is quite resistant to change, that depicts not only details that are potentially available to objective investigation by others (height, weight, hair color, gender, I.Q. score, etc)., but also items that have been learned by that person about himself or herself, either from personal experiences or by internalizing the judgments of others. A simple definition of a person's self-image is their answer to this question - "What do you believe people think about you?" A more technical term for self-image that is commonly used by social and cognitive psychologists is self-schema. Like any schema, self-schemas store information and influence the way we think and remember.
Capital punishment	Capital punishment is the killing of a person by judicial process as a punishment for an offense. Crimes that can result in a death penalty are known as capital crimes or capital offences. The term capital originates from Latin capitalis, literally "regarding the head" .
Incarceration	Incarceration is the detention of a person in jail, typically as punishment for a crime. People are most commonly incarcerated upon suspicion or conviction of committing a crime, and different jurisdictions have differing laws governing the function of incarceration within a larger system of justice. Incarceration serves four essential purposes with regard to criminals: 1. to punish criminals for committing crimes 2. to isolate criminals to prevent them from committing more crimes 3. to deter others from committing crimes 4. to rehabilitate criminals Incarceration rates, when measured by the United Nations, are considered distinct and separate from the imprisonment of political prisoners and others not charged with a specific crime.

Residual	Loosely speaking, a residual is the error in a result. To be precise, suppose we want to find x such that
	$$f(x) = b.$$
	Given an approximation x_0 of x, the residual is
	$$b - f(x_0)$$
	whereas the error is
	$$x_0 - x.$$
	If we do not know x, we cannot compute the error but we can compute the residual.
	Residual of the approximation of a function
	Similar terminology is used dealing with differential, integral, functional equations.
Poverty	Poverty is the lack of basic human needs, such as clean water, nutrition, health care, education, clothing and shelter, because of the inability to afford them. This is also referred to as absolute poverty or destitution. Relative poverty is the condition of having fewer resources or less income than others within a society or country, or compared to worldwide averages.
Unemployment	Unemployment, as defined by the International Labor Organization, occurs when people are without jobs and they have actively looked for work within the past four weeks. The unemployment rate is a measure of the prevalence of unemployment and it is calculated as a percentage by dividing the number of unemployed individuals by all individuals currently in the labor force.

Chapter 2. CLASSICAL CRIMINOLOGY

Victimisation	Victimisation is the process of being victimised. To victimize is to (1) make someone a victim or sacrifice (2) punish someone unjustly, or (3) swindle or defraud someone.
Bias	In statistics, Bias is systematic favoritism that is present in the data collection process resulting in misleading results. There are several types of statistical Bias:

· Selection Bias, where there is an error in choosing the individuals or groups to take part in a scientific study. It includes sampling Bias, in which some members of the population are more likely to be included than others. Spectrum Bias consists of evaluating the ability of a diagnostic test in a Biased group of patients, which leads to an overestimate of the sensitivity or specificity of the test. Funding Bias may lead to selection of outcomes, test samples, or test procedures that favor a study's financial sponsor.

· The Bias of an estimator is the difference between an estimator's expectation and the true value of the parameter being estimated. Omitted-variable Bias is the Bias that appears in estimates of parameters in a regression analysis when the assumed specification is incorrect, in that it omits an independent variable that should be in the model.

· In statistical hypothesis testing, a test is said to be unBiased when the probability of rejecting the null hypothesis exceeds the significance level when the alternative is true and is less than or equal to the significance level when the null hypothesis is true.

· Systematic Bias or systemic Bias are external influences that may affect the accuracy of statistical measurements.

· Data-snooping Bias comes from the misuse of data mining techniques.

Cultural bias	Cultural bias is the phenomenon of interpreting and judging phenomena by standards inherent to one's own culture. The phenomenon is sometimes considered a problem central to social and human sciences, such as economics, psychology, anthropology, and sociology. Some practitioners of the aforementioned fields have attempted to develop methods and theories to compensate for or eliminate cultural bias.

Chapter 2. CLASSICAL CRIMINOLOGY

Homicide	Homicide refers to the act of a human killing another human. A common form of homicide, for example, would be murder. It can also describe a person who has committed such an act, though this use is rare in modern English. Homicide is not always a punishable act under the criminal law, and is different than a murder from such formal legal point of view.
Sampling	Sampling is that part of statistical practice concerned with the selection of a subset of individual observations within a population of individuals intended to yield some knowledge about the population of concern, especially for the purposes of making predictions based on statistical inference. Sampling is an important aspect of data collection.
	Researchers rarely survey the entire population for two reasons (Adèr, Mellenbergh, ' Hand, 2008): the cost is too high, and the population is dynamic in that the individuals making up the population may change over time.
Sampling error	In statistics, sampling error is the error caused by observing a sample instead of the whole population. The sampling error can be found by subtracting the value of a parameter from the value of a statistic. In nursing research, a sampling error is the difference between a sample statistic used to estimate a population parameter and the actual but unknown value of the parameter (Bunns ' Grove, 2009).
Abortion	Abortion is the termination of a pregnancy by the removal or expulsion of a fetus or embryo from the uterus, resulting in or caused by its death. An abortion can occur spontaneously due to complications during pregnancy or can be induced, in humans and other species. In the context of human pregnancies, an abortion induced to preserve the health of the gravida (pregnant female) is termed a therapeutic abortion, while an abortion induced for any other reason is termed an elective abortion.

Bias	In statistics, Bias is systematic favoritism that is present in the data collection process resulting in misleading results. There are several types of statistical Bias: · Selection Bias, where there is an error in choosing the individuals or groups to take part in a scientific study. It includes sampling Bias, in which some members of the population are more likely to be included than others. Spectrum Bias consists of evaluating the ability of a diagnostic test in a Biased group of patients, which leads to an overestimate of the sensitivity or specificity of the test. Funding Bias may lead to selection of outcomes, test samples, or test procedures that favor a study's financial sponsor. · The Bias of an estimator is the difference between an estimator's expectation and the true value of the parameter being estimated. Omitted-variable Bias is the Bias that appears in estimates of parameters in a regression analysis when the assumed specification is incorrect, in that it omits an independent variable that should be in the model. · In statistical hypothesis testing, a test is said to be unBiased when the probability of rejecting the null hypothesis exceeds the significance level when the alternative is true and is less than or equal to the significance level when the null hypothesis is true. · Systematic Bias or systemic Bias are external influences that may affect the accuracy of statistical measurements. · Data-snooping Bias comes from the misuse of data mining techniques.
Cultural bias	Cultural bias is the phenomenon of interpreting and judging phenomena by standards inherent to one's own culture. The phenomenon is sometimes considered a problem central to social and human sciences, such as economics, psychology, anthropology, and sociology. Some practitioners of the aforementioned fields have attempted to develop methods and theories to compensate for or eliminate cultural bias.

Chapter 3. BIOLOGICAL FACTORS AND CRIMINAL BEHAVIOR

Probability	Probability is a way of expressing knowledge or belief that an event will occur or has occurred. The concept has been given an exact mathematical meaning in probability theory, which is used extensively in such areas of study as mathematics, statistics, finance, gambling, science, Artificial intelligence/Machine learning and philosophy to draw conclusions about the likelihood of potential events and the underlying mechanics of complex systems. Interpretations The word probability does not have a consistent direct definition.
Social learning	Social learning is learning that takes place at a wider scale than individual or group learning, up to a societal scale, through social interaction between peers. It may or may not lead to a change in attitudes and behaviour. More specifically, to be considered social learning, a process must: (1) demonstrate that a change in understanding has taken place in the individuals involved; (2) demonstrate that this change goes beyond the individual and becomes situated within wider social units or communities of practice; and (3) occur through social interactions and processes between actors within a social network (Reed et al., 2010).
Social learning theory	Social learning theory is the theory that people learn new behavior through observational learning of the social factors in their environment. If people observe positive, desired outcomes in the observed behavior, then they are more likely to model, imitate, and adopt the behavior themselves. Modern theory is closely associated with Julian Rotter and Albert Bandura.
Power	The Power of a statistical test is the probability that the test will reject a false null hypothesis (i.e. that it will not make a Type II error). As Power increases, the chances of a Type II error decrease. The probability of a Type II error is referred to as the false negative rate (β). Therefore Power is equal to $1 - \beta$.
Association	In quantitative research, the term "association" is often used to emphasize that a relationship being discussed is not necessarily causal .
Crime	Crime is the breach of rules or laws for which some governing authority (via mechanisms such as legal systems) can ultimately prescribe a conviction. Individual human societies may each define crime and crimes differently. While every crime violates the law, not every violation of the law counts as a crime; for example: breaches of contract and of other civil law may rank as "offences" or as "infractions".

25

Chapter 3. BIOLOGICAL FACTORS AND CRIMINAL BEHAVIOR

Differential	In calculus, a differential is traditionally an infinitesimally small change in a variable. For example, if x is a variable, then a change in the value of x is often denoted Δx (or δx when this change is considered to be small). The differential dx represents such a change, but is infinitely small.
Differential association	In criminology, Differential Association is a theory developed by Edwin Sutherland proposing that through interaction with others, individuals learn the values, attitudes, techniques, and motives for criminal behavior. The Differential Association Theory is the most talked about of the Interactionist theory of deviance. This theory focuses on how individuals learn how to become criminals, but does not concern itself with why they become criminals.
Adoption	Adoption is a process whereby a person assumes the parenting for another who is not kin and, in so doing, permanently transfers all rights and responsibilities from the original parent or parents. Unlike guardianship or other systems designed for the care of the young, adoption is intended to effect a permanent change in status and as such requires societal recognition, either through legal or religious sanction. Historically some societies have enacted specific laws governing adoption whereas others have endeavored to achieve adoption through less formal means, notably via contracts that specified inheritance rights and parental responsibilities.
Chicago School	In sociology and later criminology, the Chicago School was the first major body of works emerging during the 1920s and 1930s specialising in urban sociology, and the research into the urban environment by combining theory and ethnographic fieldwork in Chicago, now applied elsewhere. While involving scholars at several Chicago area universities, the term is often used interchangeably to refer to the University of Chicago's sociology department--one of the oldest and one of the most prestigious. Following World War II, a "Second Chicago School" arose whose members used symbolic interactionism combined with methods of field research, to create a new body of works.
Human Ecology	Human ecology is the interdisciplinary or transdisciplinary study of the relationship between humans and their natural, social, and built environments.

Chapter 3. BIOLOGICAL FACTORS AND CRIMINAL BEHAVIOR

Poverty	Poverty is the lack of basic human needs, such as clean water, nutrition, health care, education, clothing and shelter, because of the inability to afford them. This is also referred to as absolute poverty or destitution. Relative poverty is the condition of having fewer resources or less income than others within a society or country, or compared to worldwide averages.
Child abuse	Child abuse is the physical, sexual or emotional mistreatment of children. In the United States, the Centers for Disease Control and Prevention (CDC) define child maltreatment as any act or series of acts of commission or omission by a parent or other caregiver that results in harm, potential for harm, or threat of harm to a child. Most child abuse occurs in a child's home, with a smaller amount occurring in the organizations, schools or communities the child interacts with.
Conductance	In graph theory the conductance of a graph G=(V,E) measures how "well-knit" the graph is: it controls how fast a random walk on G converges to a uniform distribution. The conductance of a graph is often called the Cheeger constant of a graph as the analog of its counterpart in spectral geometry. Since electrical networks are intimately related to random walks with a long history in the usage of the term "conductance", this alternative name helps avoid possible confusion.
Conflict theories	Conflict theories are perspectives in social science which emphasize the social, political or material inequality of a social group, which critique the broad socio-political system, or which otherwise detract from structural functionalism and ideological conservativism. Conflict theories draw attention to power differentials, such as class conflict, and generally contrast historically dominant ideologies. Certain conflict theories set out to highlight the ideological aspects inherent in traditional thought.
Group	In the social sciences a group can be defined as two or more humans who interact with one another, accept expectations and obligations as members of the group, and share a common identity. By this definition, society can be viewed as a large group, though most social groups are considerably smaller.

29

Chapter 3. BIOLOGICAL FACTORS AND CRIMINAL BEHAVIOR

A true group exhibits some degree of social cohesion and is more than a simple collection or aggregate of individuals, such as people waiting at a bus stop.

Group conflict

Group conflicts is where social behavior causes groups of individuals to conflict with each other. It can also refer to a conflict within these groups. This conflict is often caused by differences in social norms, values, and religion.

Socialization

In economic discourse, socialization refers to the process of transforming an activity into a social relationship. Socialization of production and labor is a phenomenon that takes place under capitalism due to centralization of capital and in industries where there are increasing returns to scale, eventually leading to a situation where socialization of output (or surplus value) and co-operative ownership of the means of production is necessitated. Socialization of surplus output (profit) and ownership is one aspect of transitioning from capitalism to socialism.

Conditioning

Conditioning on the discrete level

Example. A fair coin is tossed 10 times; the random variable X is the number of heads in these 10 tosses, and Y -- the number of heads in the first 3 tosses. In spite of the fact that Y emerges before X it may happen that someone knows X but not Y.

Conditional probability

Given that X = 1, the conditional probability of the event Y = 0 is
P (Y = 0 X = 1) = P (Y = 0, X = 1) / P (X = 1) = 0.7. More generally,

$$\mathbb{P}(Y = 0 | X = x) = \frac{\binom{7}{x}}{\binom{10}{x}} = \frac{7!(10-x)!}{(7-x)!10!}$$

for x = 0, 1, 2, 3, 4, 5, 6, 7; otherwise (for x = 8, 9, 10), P (Y = 0 X = x) = 0. One may also treat the conditional probability as a random variable, -- a function of the random variable X, namely,

$$\mathbb{P}(Y = 0 | X) = \begin{cases} \binom{7}{X} / \binom{10}{X} & \text{for } X \leq 7, \\ 0 & \text{for } X > 7. \end{cases}$$

The expectation of this random variable is equal to the (unconditional) probability,

$$\mathbb{E}(\mathbb{P}(Y=0|X)) = \sum_{x} \mathbb{P}(Y=0|X=x)\mathbb{P}(X=x) = \mathbb{P}(Y=0),$$

namely,

$$\sum_{x=0}^{7} \frac{\binom{7}{x}}{\binom{10}{x}} \cdot \frac{1}{2^{10}} \binom{10}{x} = \frac{1}{8},$$

which is an instance of the law of total probability E (P (A X)) = P (A).

Thus, P (Y = 0 X = 1) may be treated as the value of the random variable P (Y = 0 X) corresponding to X = 1. *On the other hand, P (Y = 0 X = 1) is well-defined irrespective of other possible values of X.*

Conditional expectation

Given that X = 1, the conditional expectation of the random variable Y is E (Y X = 1) = 0.3. More generally,

$$\mathbb{E}(Y|X=x) = \frac{3}{10}x$$

for x = 0, .. 10. (In this example it appears to be a linear function, but in general it is nonlinear).

Cognition

Cognition is the scientific term for "the process of thought". Usage of the term varies in different disciplines; for example in psychology and cognitive science, it usually refers to an information processing view of an individual's psychological functions. Other interpretations of the meaning of cognition link it to the development of concepts; individual minds, groups, and organizations.

Chapter 3. BIOLOGICAL FACTORS AND CRIMINAL BEHAVIOR

Prejudice

A prejudice is a prejudgment, an assumption made about someone or something before having adequate knowledge to be able to do so with guaranteed accuracy. The word prejudice is most commonly used to refer to a preconceived judgment toward a people or a person because of race, social class, gender, ethnicity, age, disability, political beliefs, religion, sexual orientation or other personal characteristics. It also means beliefs without knowledge of the facts and may include "any unreasonable attitude that is unusually resistant to rational influence."

Convict

A convict is "a person found guilty of a crime and sentenced by a court" or "a person serving a sentence in prison", sometimes referred to in slang as simply a "con". Convicts are often called prisoners or inmates. Persons convicted and sentenced to non-custodial sentences often are not termed "convicts". Ex-convict (or short: ex-con) is a common way of referring to a person who has been released from prison.

Chapter 4. PSYCHOLOGICAL FACTORS AND CRIMINAL BEHAVIOR

Bias	In statistics, Bias is systematic favoritism that is present in the data collection process resulting in misleading results. There are several types of statistical Bias: · Selection Bias, where there is an error in choosing the individuals or groups to take part in a scientific study. It includes sampling Bias, in which some members of the population are more likely to be included than others. Spectrum Bias consists of evaluating the ability of a diagnostic test in a Biased group of patients, which leads to an overestimate of the sensitivity or specificity of the test. Funding Bias may lead to selection of outcomes, test samples, or test procedures that favor a study's financial sponsor. · The Bias of an estimator is the difference between an estimator's expectation and the true value of the parameter being estimated. Omitted-variable Bias is the Bias that appears in estimates of parameters in a regression analysis when the assumed specification is incorrect, in that it omits an independent variable that should be in the model. · In statistical hypothesis testing, a test is said to be unBiased when the probability of rejecting the null hypothesis exceeds the significance level when the alternative is true and is less than or equal to the significance level when the null hypothesis is true. · Systematic Bias or systemic Bias are external influences that may affect the accuracy of statistical measurements. · Data-snooping Bias comes from the misuse of data mining techniques.
Crime	Crime is the breach of rules or laws for which some governing authority (via mechanisms such as legal systems) can ultimately prescribe a conviction. Individual human societies may each define crime and crimes differently. While every crime violates the law, not every violation of the law counts as a crime; for example: breaches of contract and of other civil law may rank as "offences" or as "infractions".
Cultural bias	Cultural bias is the phenomenon of interpreting and judging phenomena by standards inherent to one's own culture. The phenomenon is sometimes considered a problem central to social and human sciences, such as economics, psychology, anthropology, and sociology. Some practitioners of the aforementioned fields have attempted to develop methods and theories to compensate for or eliminate cultural bias.

Chapter 4. PSYCHOLOGICAL FACTORS AND CRIMINAL BEHAVIOR

Deviance	In statistics, deviance is a quality of fit statistic for a model that is often used for statistical hypothesis testing. The deviance for a model M_0 is defined as $$D(y) = -2[\log\{p(y\vert\hat{\theta}_0)\} - \log\{p(y\vert\hat{\theta}_s)\}].$$
Race	Race refers to classifications of humans into large and relatively distinct populations or groups often based on factors such as appearance based on heritable phenotypical characteristics or geographic ancestry, but also often influenced by and correlated with traits such as culture, ethnicity and socio-economic status. As a biological term, race denotes genetically divergent human populations that can be marked by common phenotypic traits. This sense of race is often used by forensic anthropologists when analyzing skeletal remains, in biomedical research, and in race-based medicine.
Chicago School	In sociology and later criminology, the Chicago School was the first major body of works emerging during the 1920s and 1930s specialising in urban sociology, and the research into the urban environment by combining theory and ethnographic fieldwork in Chicago, now applied elsewhere. While involving scholars at several Chicago area universities, the term is often used interchangeably to refer to the University of Chicago's sociology department--one of the oldest and one of the most prestigious. Following World War II, a "Second Chicago School" arose whose members used symbolic interactionism combined with methods of field research, to create a new body of works.
Human Ecology	Human ecology is the interdisciplinary or transdisciplinary study of the relationship between humans and their natural, social, and built environments.
Cognition	Cognition is the scientific term for "the process of thought". Usage of the term varies in different disciplines; for example in psychology and cognitive science, it usually refers to an information processing view of an individual's psychological functions. Other interpretations of the meaning of cognition link it to the development of concepts; individual minds, groups, and organizations.
Theory	The word theory, when used by scientists, refers to an explanation of reality that has been thoroughly tested so that most scientists agree on it. It can be changed if new information is found. Theory is different from a working hypothesis, which is a theory that hasn't been fully tested; that is, a hypothesis is an unproven theory.

The word theory also distinguishes ideas from practice.

Probability

Probability is a way of expressing knowledge or belief that an event will occur or has occurred. The concept has been given an exact mathematical meaning in probability theory, which is used extensively in such areas of study as mathematics, statistics, finance, gambling, science, Artificial intelligence/Machine learning and philosophy to draw conclusions about the likelihood of potential events and the underlying mechanics of complex systems.

Interpretations

The word probability does not have a consistent direct definition.

Risk factor

In Information security, Risk factor is a collectively name for circumstances affecting the likelihood or the impact of a security risk.

Mass media

Mass media refers collectively to all media technologies, including the Internet, television, newspapers, and radio, which are used for mass communications, and to the organizations which control these technologies.

Mass media play a significant role in shaping public perceptions on a variety of important issues, both through the information that is dispensed through them, and through the interpretations they place upon this information. The also play a large role in shaping modern culture, by selecting and portraying a particular set of beliefs, values, and traditions (an entire way of life), as reality.

Social learning

Social learning is learning that takes place at a wider scale than individual or group learning, up to a societal scale, through social interaction between peers. It may or may not lead to a change in attitudes and behaviour. More specifically, to be considered social learning, a process must: (1) demonstrate that a change in understanding has taken place in the individuals involved; (2) demonstrate that this change goes beyond the individual and becomes situated within wider social units or communities of practice; and (3) occur through social interactions and processes between actors within a social network (Reed et al., 2010).

Chapter 4. PSYCHOLOGICAL FACTORS AND CRIMINAL BEHAVIOR

Social learning theory	Social learning theory is the theory that people learn new behavior through observational learning of the social factors in their environment. If people observe positive, desired outcomes in the observed behavior, then they are more likely to model, imitate, and adopt the behavior themselves. Modern theory is closely associated with Julian Rotter and Albert Bandura.
Child abuse	Child abuse is the physical, sexual or emotional mistreatment of children. In the United States, the Centers for Disease Control and Prevention (CDC) define child maltreatment as any act or series of acts of commission or omission by a parent or other caregiver that results in harm, potential for harm, or threat of harm to a child. Most child abuse occurs in a child's home, with a smaller amount occurring in the organizations, schools or communities the child interacts with.

Chapter 5. CRIME AND POVERTY

Poverty	Poverty is the lack of basic human needs, such as clean water, nutrition, health care, education, clothing and shelter, because of the inability to afford them. This is also referred to as absolute poverty or destitution. Relative poverty is the condition of having fewer resources or less income than others within a society or country, or compared to worldwide averages.
Wealth	Wealth is the abundance of valuable resources or material possessions, or the control of such assets. The word wealth is derived from the old English wela, which is from an Indo-European word stem. An individual, community, region or country that possesses an abundance of such possessions or resources is known as wealthy.
Association	In quantitative research, the term "association" is often used to emphasize that a relationship being discussed is not necessarily causal .
Differential	In calculus, a differential is traditionally an infinitesimally small change in a variable. For example, if x is a variable, then a change in the value of x is often denoted Δx (or δx when this change is considered to be small). The differential dx represents such a change, but is infinitely small.
Differential association	In criminology, Differential Association is a theory developed by Edwin Sutherland proposing that through interaction with others, individuals learn the values, attitudes, techniques, and motives for criminal behavior. The Differential Association Theory is the most talked about of the Interactionist theory of deviance. This theory focuses on how individuals learn how to become criminals, but does not concern itself with why they become criminals.
Theory	The word theory, when used by scientists, refers to an explanation of reality that has been thoroughly tested so that most scientists agree on it. It can be changed if new information is found. Theory is different from a working hypothesis, which is a theory that hasn't been fully tested; that is, a hypothesis is an unproven theory. The word theory also distinguishes ideas from practice.

Chapter 5. CRIME AND POVERTY

Crime

Crime is the breach of rules or laws for which some governing authority (via mechanisms such as legal systems) can ultimately prescribe a conviction. Individual human societies may each define crime and crimes differently. While every crime violates the law, not every violation of the law counts as a crime; for example: breaches of contract and of other civil law may rank as "offences" or as "infractions".

Bias

In statistics, Bias is systematic favoritism that is present in the data collection process resulting in misleading results. There are several types of statistical Bias:

· Selection Bias, where there is an error in choosing the individuals or groups to take part in a scientific study. It includes sampling Bias, in which some members of the population are more likely to be included than others. Spectrum Bias consists of evaluating the ability of a diagnostic test in a Biased group of patients, which leads to an overestimate of the sensitivity or specificity of the test. Funding Bias may lead to selection of outcomes, test samples, or test procedures that favor a study's financial sponsor.

· The Bias of an estimator is the difference between an estimator's expectation and the true value of the parameter being estimated. Omitted-variable Bias is the Bias that appears in estimates of parameters in a regression analysis when the assumed specification is incorrect, in that it omits an independent variable that should be in the model.

· In statistical hypothesis testing, a test is said to be unBiased when the probability of rejecting the null hypothesis exceeds the significance level when the alternative is true and is less than or equal to the significance level when the null hypothesis is true.

· Systematic Bias or systemic Bias are external influences that may affect the accuracy of statistical measurements.

· Data-snooping Bias comes from the misuse of data mining techniques.

Cultural bias

Cultural bias is the phenomenon of interpreting and judging phenomena by standards inherent to one's own culture. The phenomenon is sometimes considered a problem central to social and human sciences, such as economics, psychology, anthropology, and sociology. Some practitioners of the aforementioned fields have attempted to develop methods and theories to compensate for or eliminate cultural bias.

Chapter 5. CRIME AND POVERTY

Social learning	Social learning is learning that takes place at a wider scale than individual or group learning, up to a societal scale, through social interaction between peers. It may or may not lead to a change in attitudes and behaviour. More specifically, to be considered social learning, a process must: (1) demonstrate that a change in understanding has taken place in the individuals involved; (2) demonstrate that this change goes beyond the individual and becomes situated within wider social units or communities of practice; and (3) occur through social interactions and processes between actors within a social network (Reed et al., 2010).
Social learning theory	Social learning theory is the theory that people learn new behavior through observational learning of the social factors in their environment. If people observe positive, desired outcomes in the observed behavior, then they are more likely to model, imitate, and adopt the behavior themselves. Modern theory is closely associated with Julian Rotter and Albert Bandura.
Homicide	Homicide refers to the act of a human killing another human. A common form of homicide, for example, would be murder. It can also describe a person who has committed such an act, though this use is rare in modern English. Homicide is not always a punishable act under the criminal law, and is different than a murder from such formal legal point of view.
Unemployment	Unemployment, as defined by the International Labor Organization, occurs when people are without jobs and they have actively looked for work within the past four weeks. The unemployment rate is a measure of the prevalence of unemployment and it is calculated as a percentage by dividing the number of unemployed individuals by all individuals currently in the labor force.
Multicollinearity	Multicollinearity is a statistical phenomenon in which two or more predictor variables in a multiple regression model are highly correlated. In this situation the coefficient estimates may change erratically in response to small changes in the model or the data. Multicollinearity does not reduce the predictive power or reliability of the model as a whole; it only affects calculations regarding individual predictors.
Norm	Social norms are the behaviors and cues within a society or group. This sociological term has been defined as "the rules that a group uses for appropriate and inappropriate values, beliefs, attitudes and behaviors. These rules may be explicit or implicit.

Chapter 5. CRIME AND POVERTY

Chicago School	In sociology and later criminology, the Chicago School was the first major body of works emerging during the 1920s and 1930s specialising in urban sociology, and the research into the urban environment by combining theory and ethnographic fieldwork in Chicago, now applied elsewhere. While involving scholars at several Chicago area universities, the term is often used interchangeably to refer to the University of Chicago's sociology department--one of the oldest and one of the most prestigious. Following World War II, a "Second Chicago School" arose whose members used symbolic interactionism combined with methods of field research, to create a new body of works.
Human Ecology	Human ecology is the interdisciplinary or transdisciplinary study of the relationship between humans and their natural, social, and built environments.
Underclass	The term underclass is a coinage which functions as a morally neutral equivalent for what was known in the eighteenth and nineteenth centuries as the "undeserving poor". The earliest significant exponent of the term was the American sociologist and anthropologist Oscar Lewis in 1961. The underclass, according to Lewis, has "a strong present-time orientation, with little ability to delay gratification and plan for the future" (p. xxvi).
Labour economics	Labour economics seeks to understand the functioning and dynamics of the market for labour. Labour markets function through the interaction of workers and employers. Labour economics looks at the suppliers of labour services (workers), the demanders of labour services (employers), and attempts to understand the resulting pattern of wages, employment, and income.

Chapter 6. DURKHEIM, ANOMIE, AND MODERNIZATION

Norm

Social norms are the behaviors and cues within a society or group. This sociological term has been defined as "the rules that a group uses for appropriate and inappropriate values, beliefs, attitudes and behaviors. These rules may be explicit or implicit.

Social learning

Social learning is learning that takes place at a wider scale than individual or group learning, up to a societal scale, through social interaction between peers. It may or may not lead to a change in attitudes and behaviour. More specifically, to be considered social learning, a process must: (1) demonstrate that a change in understanding has taken place in the individuals involved; (2) demonstrate that this change goes beyond the individual and becomes situated within wider social units or communities of practice; and (3) occur through social interactions and processes between actors within a social network (Reed et al., 2010).

Social learning theory

Social learning theory is the theory that people learn new behavior through observational learning of the social factors in their environment. If people observe positive, desired outcomes in the observed behavior, then they are more likely to model, imitate, and adopt the behavior themselves. Modern theory is closely associated with Julian Rotter and Albert Bandura.

Bias

In statistics, Bias is systematic favoritism that is present in the data collection process resulting in misleading results. There are several types of statistical Bias:

· Selection Bias, where there is an error in choosing the individuals or groups to take part in a scientific study. It includes sampling Bias, in which some members of the population are more likely to be included than others. Spectrum Bias consists of evaluating the ability of a diagnostic test in a Biased group of patients, which leads to an overestimate of the sensitivity or specificity of the test. Funding Bias may lead to selection of outcomes, test samples, or test procedures that favor a study's financial sponsor.

· The Bias of an estimator is the difference between an estimator's expectation and the true value of the parameter being estimated. Omitted-variable Bias is the Bias that appears in estimates of parameters in a regression analysis when the assumed specification is incorrect, in that it omits an independent variable that should be in the model.

· In statistical hypothesis testing, a test is said to be unBiased when the probability of rejecting the null hypothesis exceeds the significance level when the alternative is true and is less than or equal to the significance level when the null hypothesis is true.

· Systematic Bias or systemic Bias are external influences that may affect the accuracy of statistical measurements.

· Data-snooping Bias comes from the misuse of data mining techniques.

Cultural bias	Cultural bias is the phenomenon of interpreting and judging phenomena by standards inherent to one's own culture. The phenomenon is sometimes considered a problem central to social and human sciences, such as economics, psychology, anthropology, and sociology. Some practitioners of the aforementioned fields have attempted to develop methods and theories to compensate for or eliminate cultural bias.
Deterrence	Deterrence is a theory from behavioral psychology about preventing or controlling actions or behavior through fear of punishment or retribution. This theory of criminology is shaping the criminal justice system of the United States and various other countries.
Erikson's stages of psychosocial development	Erikson's stages of psychosocial development as articulated by Erik Erikson explain eight stages through which a healthily developing human should pass from infancy to late adulthood. In each stage the person confronts, and hopefully masters, new challenges. Each stage builds on the successful completion of earlier stages.
Economic development	Economic development is the increase in the standard of living in a nation's population with sustained growth from a simple, low-income economy to a modern, high-income economy. Also, if the local quality of life could be improved, economic development would be enhanced. Its scope includes the process and policies by which a nation improves the economic, political, and social well-being of its people.
Modernization	In the social sciences, modernization refers to a model of an evolutionary transition from a 'pre-modern' or 'traditional' to a 'modern' society. The teleology of modernization is described in social evolutionism theories, existing as a template that has been generally followed by societies that have achieved modernity. While it may theoretically be possible for some societies to make the transition in entirely different ways, there have been no counterexamples provided by reliable sources.
Homicide	Homicide refers to the act of a human killing another human. A common form of homicide, for example, would be murder. It can also describe a person who has committed such an act, though this use is rare in modern English. Homicide is not always a punishable act under the criminal law, and is different than a murder from such formal legal point of view.

Chapter 6. DURKHEIM, ANOMIE, AND MODERNIZATION

Poverty	Poverty is the lack of basic human needs, such as clean water, nutrition, health care, education, clothing and shelter, because of the inability to afford them. This is also referred to as absolute poverty or destitution. Relative poverty is the condition of having fewer resources or less income than others within a society or country, or compared to worldwide averages.
Punishment	Punishment is the authoritative imposition of something negative or unpleasant on a person or animal in response to behavior deemed wrong by an individual or group. The authority may be either a group or a single person, and punishment may be carried out formally under a system of law or informally in other kinds of social settings such as within a family. Negative consequences that are not authorized or that are administered without a breach of rules are not considered to be punishment as defined here.
Unemployment	Unemployment, as defined by the International Labor Organization, occurs when people are without jobs and they have actively looked for work within the past four weeks. The unemployment rate is a measure of the prevalence of unemployment and it is calculated as a percentage by dividing the number of unemployed individuals by all individuals currently in the labor force.

57

Chapter 7. NEIGHBORHOODS AND CRIME

Chicago School	In sociology and later criminology, the Chicago School was the first major body of works emerging during the 1920s and 1930s specialising in urban sociology, and the research into the urban environment by combining theory and ethnographic fieldwork in Chicago, now applied elsewhere. While involving scholars at several Chicago area universities, the term is often used interchangeably to refer to the University of Chicago's sociology department--one of the oldest and one of the most prestigious. Following World War II, a "Second Chicago School" arose whose members used symbolic interactionism combined with methods of field research, to create a new body of works.
Emile Durkheim	Emile Durkheim (April 15, 1858 - November 15, 1917) was a French sociologist. He formally established the academic discipline and, with Karl Marx and Max Weber, is commonly cited as the principal architect of modern social science. Durkheim set up the first European department of sociology at the University of Bordeaux in 1895, publishing his Rules of the Sociological Method.
Human Ecology	Human ecology is the interdisciplinary or transdisciplinary study of the relationship between humans and their natural, social, and built environments.
Theory	The word theory, when used by scientists, refers to an explanation of reality that has been thoroughly tested so that most scientists agree on it. It can be changed if new information is found. Theory is different from a working hypothesis, which is a theory that hasn't been fully tested; that is, a hypothesis is an unproven theory. The word theory also distinguishes ideas from practice.
Poverty	Poverty is the lack of basic human needs, such as clean water, nutrition, health care, education, clothing and shelter, because of the inability to afford them. This is also referred to as absolute poverty or destitution. Relative poverty is the condition of having fewer resources or less income than others within a society or country, or compared to worldwide averages.

Chapter 7. NEIGHBORHOODS AND CRIME

Race	Race refers to classifications of humans into large and relatively distinct populations or groups often based on factors such as appearance based on heritable phenotypical characteristics or geographic ancestry, but also often influenced by and correlated with traits such as culture, ethnicity and socio-economic status. As a biological term, race denotes genetically divergent human populations that can be marked by common phenotypic traits. This sense of race is often used by forensic anthropologists when analyzing skeletal remains, in biomedical research, and in race-based medicine.
Social learning	Social learning is learning that takes place at a wider scale than individual or group learning, up to a societal scale, through social interaction between peers. It may or may not lead to a change in attitudes and behaviour. More specifically, to be considered social learning, a process must: (1) demonstrate that a change in understanding has taken place in the individuals involved; (2) demonstrate that this change goes beyond the individual and becomes situated within wider social units or communities of practice; and (3) occur through social interactions and processes between actors within a social network (Reed et al., 2010).
Social learning theory	Social learning theory is the theory that people learn new behavior through observational learning of the social factors in their environment. If people observe positive, desired outcomes in the observed behavior, then they are more likely to model, imitate, and adopt the behavior themselves. Modern theory is closely associated with Julian Rotter and Albert Bandura.
Punishment	Punishment is the authoritative imposition of something negative or unpleasant on a person or animal in response to behavior deemed wrong by an individual or group. The authority may be either a group or a single person, and punishment may be carried out formally under a system of law or informally in other kinds of social settings such as within a family. Negative consequences that are not authorized or that are administered without a breach of rules are not considered to be punishment as defined here.
Bias	In statistics, Bias is systematic favoritism that is present in the data collection process resulting in misleading results. There are several types of statistical Bias: · Selection Bias, where there is an error in choosing the individuals or groups to take part in a scientific study. It includes sampling Bias, in which some members of the population are more likely to be included than others. Spectrum Bias consists of evaluating the ability of a diagnostic test in a Biased group of patients, which leads to an overestimate of the sensitivity or specificity of the test. Funding Bias may lead to selection of outcomes, test samples, or test procedures that favor a study's financial sponsor.

· The Bias of an estimator is the difference between an estimator's expectation and the true value of the parameter being estimated. Omitted-variable Bias is the Bias that appears in estimates of parameters in a regression analysis when the assumed specification is incorrect, in that it omits an independent variable that should be in the model.

· In statistical hypothesis testing, a test is said to be unBiased when the probability of rejecting the null hypothesis exceeds the significance level when the alternative is true and is less than or equal to the significance level when the null hypothesis is true.

· Systematic Bias or systemic Bias are external influences that may affect the accuracy of statistical measurements.

· Data-snooping Bias comes from the misuse of data mining techniques.

Cultural bias	Cultural bias is the phenomenon of interpreting and judging phenomena by standards inherent to one's own culture. The phenomenon is sometimes considered a problem central to social and human sciences, such as economics, psychology, anthropology, and sociology. Some practitioners of the aforementioned fields have attempted to develop methods and theories to compensate for or eliminate cultural bias.
Norm	Social norms are the behaviors and cues within a society or group. This sociological term has been defined as "the rules that a group uses for appropriate and inappropriate values, beliefs, attitudes and behaviors. These rules may be explicit or implicit.
Power	The Power of a statistical test is the probability that the test will reject a false null hypothesis (i.e. that it will not make a Type II error). As Power increases, the chances of a Type II error decrease. The probability of a Type II error is referred to as the false negative rate (β). Therefore Power is equal to $1 - \beta$.
Social control	Social control refers generally to societal and political mechanisms or processes that regulate individual and group behavior, leading to conformity and compliance to the rules of a given society, state, or social group. Many mechanisms of social control are cross-cultural, if only in the control mechanisms used to prevent the establishment of chaos or anomie. Some theorists, such as Émile Durkheim, refer to this form of control as regulation.

Chapter 7. NEIGHBORHOODS AND CRIME

Capital punishment	Capital punishment is the killing of a person by judicial process as a punishment for an offense. Crimes that can result in a death penalty are known as capital crimes or capital offences. The term capital originates from Latin capitalis, literally "regarding the head" .
Family	In human context, a family is a group of people affiliated by consanguinity, affinity, or co-residence. In most societies it is the principal institution for the socialization of children. Extended from the human "family unit" by affinity, economy, culture, tradition, honor, and friendship are concepts of family that are metaphorical, or that grow increasingly inclusive extending to nationhood and humanism.
Crime	Crime is the breach of rules or laws for which some governing authority (via mechanisms such as legal systems) can ultimately prescribe a conviction. Individual human societies may each define crime and crimes differently. While every crime violates the law, not every violation of the law counts as a crime; for example: breaches of contract and of other civil law may rank as "offences" or as "infractions".
Homicide	Homicide refers to the act of a human killing another human. A common form of homicide, for example, would be murder. It can also describe a person who has committed such an act, though this use is rare in modern English. Homicide is not always a punishable act under the criminal law, and is different than a murder from such formal legal point of view.

Chapter 8. STRAIN THEORIES

Bias	In statistics, Bias is systematic favoritism that is present in the data collection process resulting in misleading results. There are several types of statistical Bias:
	· Selection Bias, where there is an error in choosing the individuals or groups to take part in a scientific study. It includes sampling Bias, in which some members of the population are more likely to be included than others. Spectrum Bias consists of evaluating the ability of a diagnostic test in a Biased group of patients, which leads to an overestimate of the sensitivity or specificity of the test. Funding Bias may lead to selection of outcomes, test samples, or test procedures that favor a study's financial sponsor.
	· The Bias of an estimator is the difference between an estimator's expectation and the true value of the parameter being estimated. Omitted-variable Bias is the Bias that appears in estimates of parameters in a regression analysis when the assumed specification is incorrect, in that it omits an independent variable that should be in the model.
	· In statistical hypothesis testing, a test is said to be unBiased when the probability of rejecting the null hypothesis exceeds the significance level when the alternative is true and is less than or equal to the significance level when the null hypothesis is true.
	· Systematic Bias or systemic Bias are external influences that may affect the accuracy of statistical measurements.
	· Data-snooping Bias comes from the misuse of data mining techniques.
Cultural bias	Cultural bias is the phenomenon of interpreting and judging phenomena by standards inherent to one's own culture. The phenomenon is sometimes considered a problem central to social and human sciences, such as economics, psychology, anthropology, and sociology. Some practitioners of the aforementioned fields have attempted to develop methods and theories to compensate for or eliminate cultural bias.
Theory	The word theory, when used by scientists, refers to an explanation of reality that has been thoroughly tested so that most scientists agree on it. It can be changed if new information is found. Theory is different from a working hypothesis, which is a theory that hasn't been fully tested; that is, a hypothesis is an unproven theory.
	The word theory also distinguishes ideas from practice.

Chapter 8. STRAIN THEORIES

Wealth	Wealth is the abundance of valuable resources or material possessions, or the control of such assets. The word wealth is derived from the old English wela, which is from an Indo-European word stem. An individual, community, region or country that possesses an abundance of such possessions or resources is known as wealthy.
Anomie	Anomie is a sociological term meaning "personal feeling of a lack of social norms; normlessness". It describes the breakdown of social norms and values. It was popularized by French sociologist Émile Durkheim in his influential book Suicide (1897).
Crime	Crime is the breach of rules or laws for which some governing authority (via mechanisms such as legal systems) can ultimately prescribe a conviction. Individual human societies may each define crime and crimes differently. While every crime violates the law, not every violation of the law counts as a crime; for example: breaches of contract and of other civil law may rank as "offences" or as "infractions".
Mean	In statistics, mean has two related meanings: • the arithmetic mean (and is distinguished from the geometric mean or harmonic mean). • the expected value of a random variable, which is also called the population mean. There are other statistical measures that use samples that some people confuse with averages - including 'median' and 'mode'. Other simple statistical analyses use measures of spread, such as range, interquartile range, or standard deviation. For a real-valued random variable X, the mean is the expectation of X. Note that not every probability distribution has a defined mean (or variance).
Norm	Social norms are the behaviors and cues within a society or group. This sociological term has been defined as "the rules that a group uses for appropriate and inappropriate values, beliefs, attitudes and behaviors. These rules may be explicit or implicit.
Social learning	Social learning is learning that takes place at a wider scale than individual or group learning, up to a societal scale, through social interaction between peers. It may or may not lead to a change in attitudes and behaviour. More specifically, to be considered social learning, a process must: (1) demonstrate that a change in understanding has taken place in the individuals involved; (2) demonstrate that this change goes beyond the individual and becomes situated within wider social units or communities of practice; and (3) occur through social interactions and processes between actors within a social network (Reed et al., 2010).

Chapter 8. STRAIN THEORIES

Social learning theory	Social learning theory is the theory that people learn new behavior through observational learning of the social factors in their environment. If people observe positive, desired outcomes in the observed behavior, then they are more likely to model, imitate, and adopt the behavior themselves. Modern theory is closely associated with Julian Rotter and Albert Bandura.
Chicago School	In sociology and later criminology, the Chicago School was the first major body of works emerging during the 1920s and 1930s specialising in urban sociology, and the research into the urban environment by combining theory and ethnographic fieldwork in Chicago, now applied elsewhere. While involving scholars at several Chicago area universities, the term is often used interchangeably to refer to the University of Chicago's sociology department--one of the oldest and one of the most prestigious. Following World War II, a "Second Chicago School" arose whose members used symbolic interactionism combined with methods of field research, to create a new body of works.
Human Ecology	Human ecology is the interdisciplinary or transdisciplinary study of the relationship between humans and their natural, social, and built environments.
Conformity	Conformity is the process by which an individual's attitudes, beliefs, and behaviors are influenced by what is conceived to be what other people might perceive. This influence occurs in both small groups and society as a whole, and it may be the result of subtle unconscious influences, or direct and overt social pressure. Conformity also occurs by the "implied presence" of others, or when other people are not actually present.
Conditioning	Conditioning on the discrete level Example. A fair coin is tossed 10 times; the random variable X is the number of heads in these 10 tosses, and Y -- the number of heads in the first 3 tosses. In spite of the fact that Y emerges before X it may happen that someone knows X but not Y. Conditional probability Given that X = 1, the conditional probability of the event Y = 0 is P (Y = 0 X = 1) = P (Y = 0, X = 1) / P (X = 1) = 0.7. More generally, $$\mathbb{P}(Y = 0 \mid X = x) = \frac{\binom{7}{x}}{\binom{10}{x}} = \frac{7!(10-x)!}{(7-x)!10!}$$

for x = 0, 1, 2, 3, 4, 5, 6, 7; otherwise (for x = 8, 9, 10), P (Y = 0 X = x) = 0. One may also treat the conditional probability as a random variable, -- a function of the random variable X, namely,

$$\mathbb{P}(Y = 0 | X) = \begin{cases} \binom{7}{X} / \binom{10}{X} & \text{for } X \leq 7, \\ 0 & \text{for } X > 7. \end{cases}$$

The expectation of this random variable is equal to the (unconditional) probability,

$$\mathbb{E}(\mathbb{P}(Y = 0 | X)) = \sum_x \mathbb{P}(Y = 0 | X = x) \mathbb{P}(X = x) = \mathbb{P}(Y = 0),$$

namely,

$$\sum_{x=0}^{7} \frac{\binom{7}{x}}{\binom{10}{x}} \cdot \frac{1}{2^{10}} \binom{10}{x} = \frac{1}{8},$$

which is an instance of the law of total probability E (P (A X)) = P (A).

Thus, P (Y = 0 X = 1) may be treated as the value of the random variable P (Y = 0 X) corresponding to X = 1. *On the other hand, P (Y = 0 X = 1) is well-defined irrespective of other possible values of X.*

Conditional expectation

Given that X = 1, the conditional expectation of the random variable Y is E (Y X = 1) = 0.3. More generally,

$$\mathbb{E}(Y | X = x) = \frac{3}{10} x$$

for x = 0, .. 10. (In this example it appears to be a linear function, but in general it is nonlinear).

Rebellion

Rebellion is a refusal of obedience or order. It may, therefore, be seen as encompassing a range of behaviors from civil disobedience and mass nonviolent resistance, to violent and organized attempts to destroy an established authority such as a government. Those who participate in rebellions are known as "rebels".

Social status

In sociology or anthropology, social status is the honor or prestige attached to one's position in society (one's social position). It may also refer to a rank or position that one holds in a group, for example son or daughter, playmate, pupil, etc.

Social status, the position or rank of a person or group within the society, can be determined two ways.

Homicide

Homicide refers to the act of a human killing another human. A common form of homicide, for example, would be murder. It can also describe a person who has committed such an act, though this use is rare in modern English. Homicide is not always a punishable act under the criminal law, and is different than a murder from such formal legal point of view.

Subculture

In sociology, anthropology and cultural studies, a subculture is a group of people with a culture (whether distinct or hidden) which differentiates them from the larger culture to which they belong.

As early as 1950, David Riesman distinguished between a majority, "which passively accepted commercially provided styles and meanings, and a 'subculture' which actively sought a minority style ... and interpreted it in accordance with subversive values".

Poverty

Poverty is the lack of basic human needs, such as clean water, nutrition, health care, education, clothing and shelter, because of the inability to afford them. This is also referred to as absolute poverty or destitution. Relative poverty is the condition of having fewer resources or less income than others within a society or country, or compared to worldwide averages.

Chapter 8. STRAIN THEORIES

Cognition	Cognition is the scientific term for "the process of thought". Usage of the term varies in different disciplines; for example in psychology and cognitive science, it usually refers to an information processing view of an individual's psychological functions. Other interpretations of the meaning of cognition link it to the development of concepts; individual minds, groups, and organizations.
Non-Hispanic Whites	Non-Hispanic Whites, Not Hispanic or Latino are people in the United States, as defined by the Census Bureau, who are of the White race but not of Hispanic or Latino origin/ethnicity. Their origins are in Europe (the vast majority), North Africa, the Middle East, and elsewhere, and date from English, French, and other European colonization that began in the 16th century, and from the massive immigration that occurred since independence in the 18th century. They are more or less equivalent to European Americans, although not exactly; "European American" in its technical definition excludes people from North Africa and the Middle East, and includes Hispanic or Latino Whites of primarily Spanish and other European background.
Strain theory	In criminology, the strain theory states that social structures within society may encourage citizens to commit crime. Following on the work of Émile Durkheim, Strain Theories have been advanced by Robert King Merton (1938), Albert K. Cohen (1955), Richard Cloward and Lloyd Ohlin (1960), Robert Agnew (1992), and Steven Messner and Richard Rosenfeld (1994). Strain may be either: • Structural: this refers to the processes at the societal level which filter down and affect how the individual perceives his or her needs, i.e. if particular social structures are inherently inadequate or there is inadequate regulation, this may change the individual's perceptions as to means and opportunities; or • Individual: this refers to the frictions and pains experienced by an individual as he or she looks for ways to satisfy his or her needs, i.e. if the goals of a society become significant to an individual, actually achieving them may become more important than the means adopted.
Probability	Probability is a way of expressing knowledge or belief that an event will occur or has occurred. The concept has been given an exact mathematical meaning in probability theory, which is used extensively in such areas of study as mathematics, statistics, finance, gambling, science, Artificial intelligence/Machine learning and philosophy to draw conclusions about the likelihood of potential events and the underlying mechanics of complex systems. Interpretations

The word probability does not have a consistent direct definition.

Social safety net

Social Safety Nets are non-contributory transfer programs seeking to prevent the poor or those vulnerable to shocks and poverty from falling below a certain poverty level. Safety net programs can be provided by the public sector (the State and aid donors) or by the private sector (NGOs, private firms, charities, and informal household transfers). Safety net transfers include:

- Cash transfers
- Food-based programs such as supplementary feeding programs and food stamps, vouchers, and coupons
- In-kind transfers such as school supplies and uniforms
- Conditional cash transfers
- Price subsidies for food, electricity, or public transport
- Public works
- Fee waivers and exemptions for health care, schooling and utilities

On average, spending on safety nets accounts for 1 to 2 percent of GDP across developing and transition countries, though sometimes much less or much more.

Social control

Social control refers generally to societal and political mechanisms or processes that regulate individual and group behavior, leading to conformity and compliance to the rules of a given society, state, or social group. Many mechanisms of social control are cross-cultural, if only in the control mechanisms used to prevent the establishment of chaos or anomie. Some theorists, such as Émile Durkheim, refer to this form of control as regulation.

Chapter 9. LEARNING THEORIES

Social learning	Social learning is learning that takes place at a wider scale than individual or group learning, up to a societal scale, through social interaction between peers. It may or may not lead to a change in attitudes and behaviour. More specifically, to be considered social learning, a process must: (1) demonstrate that a change in understanding has taken place in the individuals involved; (2) demonstrate that this change goes beyond the individual and becomes situated within wider social units or communities of practice; and (3) occur through social interactions and processes between actors within a social network (Reed et al., 2010).
Social learning theory	Social learning theory is the theory that people learn new behavior through observational learning of the social factors in their environment. If people observe positive, desired outcomes in the observed behavior, then they are more likely to model, imitate, and adopt the behavior themselves. Modern theory is closely associated with Julian Rotter and Albert Bandura.
Chicago School	In sociology and later criminology, the Chicago School was the first major body of works emerging during the 1920s and 1930s specialising in urban sociology, and the research into the urban environment by combining theory and ethnographic fieldwork in Chicago, now applied elsewhere. While involving scholars at several Chicago area universities, the term is often used interchangeably to refer to the University of Chicago's sociology department--one of the oldest and one of the most prestigious. Following World War II, a "Second Chicago School" arose whose members used symbolic interactionism combined with methods of field research, to create a new body of works.
Human Ecology	Human ecology is the interdisciplinary or transdisciplinary study of the relationship between humans and their natural, social, and built environments.
Cognition	Cognition is the scientific term for "the process of thought". Usage of the term varies in different disciplines; for example in psychology and cognitive science, it usually refers to an information processing view of an individual's psychological functions. Other interpretations of the meaning of cognition link it to the development of concepts; individual minds, groups, and organizations.
Conditioning	Conditioning on the discrete level

Example. A fair coin is tossed 10 times; the random variable X is the number of heads in these 10 tosses, and Y -- the number of heads in the first 3 tosses. In spite of the fact that Y emerges before X it may happen that someone knows X but not Y.

Conditional probability |

Given that X = 1, the conditional probability of the event Y = 0 is
P (Y = 0 X = 1) = P (Y = 0, X = 1) / P (X = 1) = 0.7. More generally,

$$\mathbb{P}(Y = 0 | X = x) = \frac{\binom{7}{x}}{\binom{10}{x}} = \frac{7!(10 - x)!}{(7 - x)!10!}$$

for x = 0, 1, 2, 3, 4, 5, 6, 7; otherwise (for x = 8, 9, 10), P (Y = 0 X = x) = 0. One may also treat the conditional probability as a random variable, -- a function of the random variable X, namely,

$$\mathbb{P}(Y = 0 | X) = \begin{cases} \binom{7}{X}/\binom{10}{X} & \text{for } X \leq 7, \\ 0 & \text{for } X > 7. \end{cases}$$

The expectation of this random variable is equal to the (unconditional) probability,

$$\mathbb{E}(\mathbb{P}(Y = 0 | X)) = \sum_{x} \mathbb{P}(Y = 0 | X = x)\mathbb{P}(X = x) = \mathbb{P}(Y = 0),$$

namely,

$$\sum_{x=0}^{7} \frac{\binom{7}{x}}{\binom{10}{x}} \cdot \frac{1}{2^{10}} \binom{10}{x} = \frac{1}{8},$$

which is an instance of the law of total probability E (P (A X)) = P (A).

Thus, P (Y = 0 X = 1) may be treated as the value of the random variable P (Y = 0 X) corresponding to X = 1. *On the other hand, P (Y = 0 X = 1) is well-defined irrespective of other possible values of X.*

Conditional expectation

Given that X = 1, the conditional expectation of the random variable Y is E (Y X = 1) = 0.3. More generally,

$$\mathbb{E}(Y|X = x) = \frac{3}{10}x$$

for x = 0, .. 10. (In this example it appears to be a linear function, but in general it is nonlinear).

Punishment	Punishment is the authoritative imposition of something negative or unpleasant on a person or animal in response to behavior deemed wrong by an individual or group. The authority may be either a group or a single person, and punishment may be carried out formally under a system of law or informally in other kinds of social settings such as within a family. Negative consequences that are not authorized or that are administered without a breach of rules are not considered to be punishment as defined here.
Association	In quantitative research, the term "association" is often used to emphasize that a relationship being discussed is not necessarily causal .
Differential	In calculus, a differential is traditionally an infinitesimally small change in a variable. For example, if x is a variable, then a change in the value of x is often denoted Δx (or δx when this change is considered to be small). The differential dx represents such a change, but is infinitely small.
Differential association	In criminology, Differential Association is a theory developed by Edwin Sutherland proposing that through interaction with others, individuals learn the values, attitudes, techniques, and motives for criminal behavior.
	The Differential Association Theory is the most talked about of the Interactionist theory of deviance. This theory focuses on how individuals learn how to become criminals, but does not concern itself with why they become criminals.

Chapter 9. LEARNING THEORIES

Crime	Crime is the breach of rules or laws for which some governing authority (via mechanisms such as legal systems) can ultimately prescribe a conviction. Individual human societies may each define crime and crimes differently. While every crime violates the law, not every violation of the law counts as a crime; for example: breaches of contract and of other civil law may rank as "offences" or as "infractions".
Interactionism	In sociology, interactionism is a theoretical perspective that derives social processes (such as conflict, cooperation, identity formation) from human interaction. It is the study of individuals and how they act within society.
Symbolic interactionism	Symbolic interactionism is a major sociological perspective that places emphasis on micro-scale social interaction, which is particularly important in subfields such as urban sociology and social psychology. Symbolic interactionism is derived from American pragmatism, especially the work of George Herbert Mead and Charles Cooley. Herbert Blumer, a student and interpreter of Mead, coined the term and put forward an influential summary of the perspective: people act toward things based on the meaning those things have for them; and these meanings are derived from social interaction and modified through interpretation.
Bias	In statistics, Bias is systematic favoritism that is present in the data collection process resulting in misleading results. There are several types of statistical Bias:
	· Selection Bias, where there is an error in choosing the individuals or groups to take part in a scientific study. It includes sampling Bias, in which some members of the population are more likely to be included than others. Spectrum Bias consists of evaluating the ability of a diagnostic test in a Biased group of patients, which leads to an overestimate of the sensitivity or specificity of the test. Funding Bias may lead to selection of outcomes, test samples, or test procedures that favor a study's financial sponsor.
	· The Bias of an estimator is the difference between an estimator's expectation and the true value of the parameter being estimated. Omitted-variable Bias is the Bias that appears in estimates of parameters in a regression analysis when the assumed specification is incorrect, in that it omits an independent variable that should be in the model.
	· In statistical hypothesis testing, a test is said to be unBiased when the probability of rejecting the null hypothesis exceeds the significance level when the alternative is true and is less than or equal to the significance level when the null hypothesis is true.
	· Systematic Bias or systemic Bias are external influences that may affect the accuracy of statistical measurements.

Chapter 9. LEARNING THEORIES

· Data-snooping Bias comes from the misuse of data mining techniques.

Cultural bias	Cultural bias is the phenomenon of interpreting and judging phenomena by standards inherent to one's own culture. The phenomenon is sometimes considered a problem central to social and human sciences, such as economics, psychology, anthropology, and sociology. Some practitioners of the aforementioned fields have attempted to develop methods and theories to compensate for or eliminate cultural bias.
Poverty	Poverty is the lack of basic human needs, such as clean water, nutrition, health care, education, clothing and shelter, because of the inability to afford them. This is also referred to as absolute poverty or destitution. Relative poverty is the condition of having fewer resources or less income than others within a society or country, or compared to worldwide averages.
Homicide	Homicide refers to the act of a human killing another human. A common form of homicide, for example, would be murder. It can also describe a person who has committed such an act, though this use is rare in modern English. Homicide is not always a punishable act under the criminal law, and is different than a murder from such formal legal point of view.
Self-image	A person's self-image is the mental picture, generally of a kind that is quite resistant to change, that depicts not only details that are potentially available to objective investigation by others (height, weight, hair color, gender, I.Q. score, etc)., but also items that have been learned by that person about himself or herself, either from personal experiences or by internalizing the judgments of others. A simple definition of a person's self-image is their answer to this question - "What do you believe people think about you?" A more technical term for self-image that is commonly used by social and cognitive psychologists is self-schema. Like any schema, self-schemas store information and influence the way we think and remember.
Subculture	In sociology, anthropology and cultural studies, a subculture is a group of people with a culture (whether distinct or hidden) which differentiates them from the larger culture to which they belong. As early as 1950, David Riesman distinguished between a majority, "which passively accepted commercially provided styles and meanings, and a 'subculture' which actively sought a minority style ... and interpreted it in accordance with subversive values".

Chapter 9. LEARNING THEORIES

Labelling	Labelling is describing someone or something in a word or short phrase. For example, describing someone who has broken a law as a criminal. Labeling theory is a theory in sociology which ascribes labelling of people to control and identification of deviant behavior.
Labour economics	Labour economics seeks to understand the functioning and dynamics of the market for labour. Labour markets function through the interaction of workers and employers. Labour economics looks at the suppliers of labour services (workers), the demanders of labour services (employers), and attempts to understand the resulting pattern of wages, employment, and income.
Non-Hispanic Whites	Non-Hispanic Whites, Not Hispanic or Latino are people in the United States, as defined by the Census Bureau, who are of the White race but not of Hispanic or Latino origin/ethnicity. Their origins are in Europe (the vast majority), North Africa, the Middle East, and elsewhere, and date from English, French, and other European colonization that began in the 16th century, and from the massive immigration that occurred since independence in the 18th century. They are more or less equivalent to European Americans, although not exactly; "European American" in its technical definition excludes people from North Africa and the Middle East, and includes Hispanic or Latino Whites of primarily Spanish and other European background.

Chapter 10. CONTROL THEORIES

Theory	The word theory, when used by scientists, refers to an explanation of reality that has been thoroughly tested so that most scientists agree on it. It can be changed if new information is found. Theory is different from a working hypothesis, which is a theory that hasn't been fully tested; that is, a hypothesis is an unproven theory. The word theory also distinguishes ideas from practice.
Conformity	Conformity is the process by which an individual's attitudes, beliefs, and behaviors are influenced by what is conceived to be what other people might perceive. This influence occurs in both small groups and society as a whole, and it may be the result of subtle unconscious influences, or direct and overt social pressure. Conformity also occurs by the "implied presence" of others, or when other people are not actually present.
Norm	Social norms are the behaviors and cues within a society or group. This sociological term has been defined as "the rules that a group uses for appropriate and inappropriate values, beliefs, attitudes and behaviors. These rules may be explicit or implicit.
Punishment	Punishment is the authoritative imposition of something negative or unpleasant on a person or animal in response to behavior deemed wrong by an individual or group. The authority may be either a group or a single person, and punishment may be carried out formally under a system of law or informally in other kinds of social settings such as within a family. Negative consequences that are not authorized or that are administered without a breach of rules are not considered to be punishment as defined here.
Social control	Social control refers generally to societal and political mechanisms or processes that regulate individual and group behavior, leading to conformity and compliance to the rules of a given society, state, or social group. Many mechanisms of social control are cross-cultural, if only in the control mechanisms used to prevent the establishment of chaos or anomie. Some theorists, such as Émile Durkheim, refer to this form of control as regulation.
Social learning	Social learning is learning that takes place at a wider scale than individual or group learning, up to a societal scale, through social interaction between peers. It may or may not lead to a change in attitudes and behaviour. More specifically, to be considered social learning, a process must: (1) demonstrate that a change in understanding has taken place in the individuals involved; (2) demonstrate that this change goes beyond the individual and becomes situated within wider social units or communities of practice; and (3) occur through social interactions and processes between actors within a social network (Reed et al., 2010).

Chapter 10. CONTROL THEORIES

Social learning theory	Social learning theory is the theory that people learn new behavior through observational learning of the social factors in their environment. If people observe positive, desired outcomes in the observed behavior, then they are more likely to model, imitate, and adopt the behavior themselves. Modern theory is closely associated with Julian Rotter and Albert Bandura.
Socialization	In economic discourse, socialization refers to the process of transforming an activity into a social relationship. Socialization of production and labor is a phenomenon that takes place under capitalism due to centralization of capital and in industries where there are increasing returns to scale, eventually leading to a situation where socialization of output (or surplus value) and co-operative ownership of the means of production is necessitated. Socialization of surplus output (profit) and ownership is one aspect of transitioning from capitalism to socialism.
Chicago School	In sociology and later criminology, the Chicago School was the first major body of works emerging during the 1920s and 1930s specialising in urban sociology, and the research into the urban environment by combining theory and ethnographic fieldwork in Chicago, now applied elsewhere. While involving scholars at several Chicago area universities, the term is often used interchangeably to refer to the University of Chicago's sociology department--one of the oldest and one of the most prestigious. Following World War II, a "Second Chicago School" arose whose members used symbolic interactionism combined with methods of field research, to create a new body of works.
Human Ecology	Human ecology is the interdisciplinary or transdisciplinary study of the relationship between humans and their natural, social, and built environments.
Association	In quantitative research, the term "association" is often used to emphasize that a relationship being discussed is not necessarily causal .
Conflict theories	Conflict theories are perspectives in social science which emphasize the social, political or material inequality of a social group, which critique the broad socio-political system, or which otherwise detract from structural functionalism and ideological conservativism. Conflict theories draw attention to power differentials, such as class conflict, and generally contrast historically dominant ideologies. Certain conflict theories set out to highlight the ideological aspects inherent in traditional thought.

Chapter 10. CONTROL THEORIES

Crime	Crime is the breach of rules or laws for which some governing authority (via mechanisms such as legal systems) can ultimately prescribe a conviction. Individual human societies may each define crime and crimes differently. While every crime violates the law, not every violation of the law counts as a crime; for example: breaches of contract and of other civil law may rank as "offences" or as "infractions".
Differential	In calculus, a differential is traditionally an infinitesimally small change in a variable. For example, if x is a variable, then a change in the value of x is often denoted Δx (or δx when this change is considered to be small). The differential dx represents such a change, but is infinitely small.
Differential association	In criminology, Differential Association is a theory developed by Edwin Sutherland proposing that through interaction with others, individuals learn the values, attitudes, techniques, and motives for criminal behavior. The Differential Association Theory is the most talked about of the Interactionist theory of deviance. This theory focuses on how individuals learn how to become criminals, but does not concern itself with why they become criminals.
Group	In the social sciences a group can be defined as two or more humans who interact with one another, accept expectations and obligations as members of the group, and share a common identity. By this definition, society can be viewed as a large group, though most social groups are considerably smaller. A true group exhibits some degree of social cohesion and is more than a simple collection or aggregate of individuals, such as people waiting at a bus stop.
Group conflict	Group conflicts is where social behavior causes groups of individuals to conflict with each other. It can also refer to a conflict within these groups. This conflict is often caused by differences in social norms, values, and religion.

Chapter 10. CONTROL THEORIES

Cognition	Cognition is the scientific term for "the process of thought". Usage of the term varies in different disciplines; for example in psychology and cognitive science, it usually refers to an information processing view of an individual's psychological functions. Other interpretations of the meaning of cognition link it to the development of concepts; individual minds, groups, and organizations.
Bias	In statistics, Bias is systematic favoritism that is present in the data collection process resulting in misleading results. There are several types of statistical Bias:
	· Selection Bias, where there is an error in choosing the individuals or groups to take part in a scientific study. It includes sampling Bias, in which some members of the population are more likely to be included than others. Spectrum Bias consists of evaluating the ability of a diagnostic test in a Biased group of patients, which leads to an overestimate of the sensitivity or specificity of the test. Funding Bias may lead to selection of outcomes, test samples, or test procedures that favor a study's financial sponsor.
	· The Bias of an estimator is the difference between an estimator's expectation and the true value of the parameter being estimated. Omitted-variable Bias is the Bias that appears in estimates of parameters in a regression analysis when the assumed specification is incorrect, in that it omits an independent variable that should be in the model.
	· In statistical hypothesis testing, a test is said to be unBiased when the probability of rejecting the null hypothesis exceeds the significance level when the alternative is true and is less than or equal to the significance level when the null hypothesis is true.
	· Systematic Bias or systemic Bias are external influences that may affect the accuracy of statistical measurements.
	· Data-snooping Bias comes from the misuse of data mining techniques.
Cultural bias	Cultural bias is the phenomenon of interpreting and judging phenomena by standards inherent to one's own culture. The phenomenon is sometimes considered a problem central to social and human sciences, such as economics, psychology, anthropology, and sociology. Some practitioners of the aforementioned fields have attempted to develop methods and theories to compensate for or eliminate cultural bias.

Chapter 10. CONTROL THEORIES

Poverty	Poverty is the lack of basic human needs, such as clean water, nutrition, health care, education, clothing and shelter, because of the inability to afford them. This is also referred to as absolute poverty or destitution. Relative poverty is the condition of having fewer resources or less income than others within a society or country, or compared to worldwide averages.
Unemployment	Unemployment, as defined by the International Labor Organization, occurs when people are without jobs and they have actively looked for work within the past four weeks. The unemployment rate is a measure of the prevalence of unemployment and it is calculated as a percentage by dividing the number of unemployed individuals by all individuals currently in the labor force.
Attention deficit hyperactivity disorder	Attention deficit hyperactivity disorder is a neurobehavioral developmental disorder. It is primarily characterized by "the co-existence of attentional problems and hyperactivity, with each behavior occurring infrequently alone" and symptoms starting before seven years of age. Attention deficit hyperactivity disorder is the most commonly studied and diagnosed psychiatric disorder in children, affecting about 3 to 5 percent of children globally and diagnosed in about 2 to 16 percent of school aged children.
Labelling	Labelling is describing someone or something in a word or short phrase. For example, describing someone who has broken a law as a criminal. Labeling theory is a theory in sociology which ascribes labelling of people to control and identification of deviant behavior.
Organized crime	Organized crime are transnational, national, or local groupings of highly centralized enterprises run by criminals for the purpose of engaging in illegal activity, most commonly for monetary profit. The Organized Crime Control Act (U.S., 1970) defines organized crime as "The unlawful activities of [...] a highly organized, disciplined association [...]". Such crime is commonly referred to as the work of the Mob in the U.S.
Non-Hispanic Whites	Non-Hispanic Whites, Not Hispanic or Latino are people in the United States, as defined by the Census Bureau, who are of the White race but not of Hispanic or Latino origin/ethnicity. Their origins are in Europe (the vast majority), North Africa, the Middle East, and elsewhere, and date from English, French, and other European colonization that began in the 16th century, and from the massive immigration that occurred since independence in the 18th century. They are more or less equivalent to European Americans, although not exactly; "European American" in its technical definition excludes people from North Africa and the Middle East, and includes Hispanic or Latino Whites of primarily Spanish and other European background.

Chapter 10. CONTROL THEORIES

Subculture	In sociology, anthropology and cultural studies, a subculture is a group of people with a culture (whether distinct or hidden) which differentiates them from the larger culture to which they belong.
	As early as 1950, David Riesman distinguished between a majority, "which passively accepted commercially provided styles and meanings, and a 'subculture' which actively sought a minority style ... and interpreted it in accordance with subversive values".

Chapter 11. THE MEANING OF CRIME

Association	In quantitative research, the term "association" is often used to emphasize that a relationship being discussed is not necessarily causal .
Crime	Crime is the breach of rules or laws for which some governing authority (via mechanisms such as legal systems) can ultimately prescribe a conviction. Individual human societies may each define crime and crimes differently. While every crime violates the law, not every violation of the law counts as a crime; for example: breaches of contract and of other civil law may rank as "offences" or as "infractions".
Differential	In calculus, a differential is traditionally an infinitesimally small change in a variable. For example, if x is a variable, then a change in the value of x is often denoted Δx (or δx when this change is considered to be small). The differential dx represents such a change, but is infinitely small.
Differential association	In criminology, Differential Association is a theory developed by Edwin Sutherland proposing that through interaction with others, individuals learn the values, attitudes, techniques, and motives for criminal behavior. The Differential Association Theory is the most talked about of the Interactionist theory of deviance. This theory focuses on how individuals learn how to become criminals, but does not concern itself with why they become criminals.
Interactionism	In sociology, interactionism is a theoretical perspective that derives social processes (such as conflict, cooperation, identity formation) from human interaction. It is the study of individuals and how they act within society.
Labelling	Labelling is describing someone or something in a word or short phrase. For example, describing someone who has broken a law as a criminal. Labeling theory is a theory in sociology which ascribes labelling of people to control and identification of deviant behavior.
Labeling theory	Originating in sociology and criminology, labeling theory was developed by sociologist Howard S. Becker. Labeling theory holds that deviance is not inherent to an act, but instead focuses on the linguistic tendency of majorities to negatively label minorities or those seen as deviant from norms. The theory is concerned with how the self-identity and behavior of individuals may be determined or influenced by the terms used to describe or classify them, and is associated with the concept of a self-fulfilling prophecy and stereotyping.

Chapter 11. THE MEANING OF CRIME

Symbolic interactionism	Symbolic interactionism is a major sociological perspective that places emphasis on micro-scale social interaction, which is particularly important in subfields such as urban sociology and social psychology. Symbolic interactionism is derived from American pragmatism, especially the work of George Herbert Mead and Charles Cooley. Herbert Blumer, a student and interpreter of Mead, coined the term and put forward an influential summary of the perspective: people act toward things based on the meaning those things have for them; and these meanings are derived from social interaction and modified through interpretation.
Social control	Social control refers generally to societal and political mechanisms or processes that regulate individual and group behavior, leading to conformity and compliance to the rules of a given society, state, or social group. Many mechanisms of social control are cross-cultural, if only in the control mechanisms used to prevent the establishment of chaos or anomie. Some theorists, such as Émile Durkheim, refer to this form of control as regulation.
Homicide	Homicide refers to the act of a human killing another human. A common form of homicide, for example, would be murder. It can also describe a person who has committed such an act, though this use is rare in modern English. Homicide is not always a punishable act under the criminal law, and is different than a murder from such formal legal point of view.
Norm	Social norms are the behaviors and cues within a society or group. This sociological term has been defined as "the rules that a group uses for appropriate and inappropriate values, beliefs, attitudes and behaviors. These rules may be explicit or implicit.
Social learning	Social learning is learning that takes place at a wider scale than individual or group learning, up to a societal scale, through social interaction between peers. It may or may not lead to a change in attitudes and behaviour. More specifically, to be considered social learning, a process must: (1) demonstrate that a change in understanding has taken place in the individuals involved; (2) demonstrate that this change goes beyond the individual and becomes situated within wider social units or communities of practice; and (3) occur through social interactions and processes between actors within a social network (Reed et al., 2010).
Social learning theory	Social learning theory is the theory that people learn new behavior through observational learning of the social factors in their environment. If people observe positive, desired outcomes in the observed behavior, then they are more likely to model, imitate, and adopt the behavior themselves. Modern theory is closely associated with Julian Rotter and Albert Bandura.
Bias	In statistics, Bias is systematic favoritism that is present in the data collection process resulting in misleading results. There are several types of statistical Bias:

107

· Selection Bias, where there is an error in choosing the individuals or groups to take part in a scientific study. It includes sampling Bias, in which some members of the population are more likely to be included than others. Spectrum Bias consists of evaluating the ability of a diagnostic test in a Biased group of patients, which leads to an overestimate of the sensitivity or specificity of the test. Funding Bias may lead to selection of outcomes, test samples, or test procedures that favor a study's financial sponsor.

· The Bias of an estimator is the difference between an estimator's expectation and the true value of the parameter being estimated. Omitted-variable Bias is the Bias that appears in estimates of parameters in a regression analysis when the assumed specification is incorrect, in that it omits an independent variable that should be in the model.

· In statistical hypothesis testing, a test is said to be unBiased when the probability of rejecting the null hypothesis exceeds the significance level when the alternative is true and is less than or equal to the significance level when the null hypothesis is true.

· Systematic Bias or systemic Bias are external influences that may affect the accuracy of statistical measurements.

· Data-snooping Bias comes from the misuse of data mining techniques.

Cultural bias	Cultural bias is the phenomenon of interpreting and judging phenomena by standards inherent to one's own culture. The phenomenon is sometimes considered a problem central to social and human sciences, such as economics, psychology, anthropology, and sociology. Some practitioners of the aforementioned fields have attempted to develop methods and theories to compensate for or eliminate cultural bias.
Conditioning	Conditioning on the discrete level
	Example. A fair coin is tossed 10 times; the random variable X is the number of heads in these 10 tosses, and Y -- the number of heads in the first 3 tosses. In spite of the fact that Y emerges before X it may happen that someone knows X but not Y.
	Conditional probability

Given that X = 1, the conditional probability of the event Y = 0 is
P (Y = 0 X = 1) = P (Y = 0, X = 1) / P (X = 1) = 0.7. More generally,

$$\mathbb{P}(Y = 0 | X = x) = \frac{\binom{7}{x}}{\binom{10}{x}} = \frac{7!(10 - x)!}{(7 - x)!10!}$$

for x = 0, 1, 2, 3, 4, 5, 6, 7; otherwise (for x = 8, 9, 10), P (Y = 0 X = x) = 0. One may also treat the conditional probability as a random variable, -- a function of the random variable X, namely,

$$\mathbb{P}(Y = 0 | X) = \begin{cases} \binom{7}{X}/\binom{10}{X} & \text{for } X \leq 7, \\ 0 & \text{for } X > 7. \end{cases}$$

The expectation of this random variable is equal to the (unconditional) probability,

$$\mathbb{E}(\mathbb{P}(Y = 0 | X)) = \sum_{x} \mathbb{P}(Y = 0 | X = x)\mathbb{P}(X = x) = \mathbb{P}(Y = 0),$$

namely,

$$\sum_{x=0}^{7} \frac{\binom{7}{x}}{\binom{10}{x}} \cdot \frac{1}{2^{10}} \binom{10}{x} = \frac{1}{8},$$

which is an instance of the law of total probability E (P (A X)) = P (A).

Thus, P (Y = 0 X = 1) may be treated as the value of the random variable P (Y = 0 X) corresponding to X = 1. *On the other hand, P (Y = 0 X = 1) is well-defined irrespective of other possible values of X.*

Conditional expectation

Given that X = 1, the conditional expectation of the random variable Y is E (Y X = 1) = 0.3. More generally,

$$\mathbb{E}(Y|X=x) = \frac{3}{10}x$$

for x = 0, .. 10. (In this example it appears to be a linear function, but in general it is nonlinear).

Poverty

Poverty is the lack of basic human needs, such as clean water, nutrition, health care, education, clothing and shelter, because of the inability to afford them. This is also referred to as absolute poverty or destitution. Relative poverty is the condition of having fewer resources or less income than others within a society or country, or compared to worldwide averages.

Organized crime

Organized crime are transnational, national, or local groupings of highly centralized enterprises run by criminals for the purpose of engaging in illegal activity, most commonly for monetary profit. The Organized Crime Control Act (U.S., 1970) defines organized crime as "The unlawful activities of [...] a highly organized, disciplined association [...]". Such crime is commonly referred to as the work of the Mob in the U.S.

Chapter 12. CONFLICT CRIMINOLOGY

Theory	The word theory, when used by scientists, refers to an explanation of reality that has been thoroughly tested so that most scientists agree on it. It can be changed if new information is found. Theory is different from a working hypothesis, which is a theory that hasn't been fully tested; that is, a hypothesis is an unproven theory. The word theory also distinguishes ideas from practice.
Urbanization	Urbanization is the physical growth of urban areas as a result of global change. Urbanization is also defined by the United Nations as movement of people from rural to urban areas with population growth equating to urban migration. The United Nations projected that half of the world's population would live in urban areas at the end of 2008.
Norm	Social norms are the behaviors and cues within a society or group. This sociological term has been defined as "the rules that a group uses for appropriate and inappropriate values, beliefs, attitudes and behaviors. These rules may be explicit or implicit.
Social learning	Social learning is learning that takes place at a wider scale than individual or group learning, up to a societal scale, through social interaction between peers. It may or may not lead to a change in attitudes and behaviour. More specifically, to be considered social learning, a process must: (1) demonstrate that a change in understanding has taken place in the individuals involved; (2) demonstrate that this change goes beyond the individual and becomes situated within wider social units or communities of practice; and (3) occur through social interactions and processes between actors within a social network (Reed et al., 2010).
Social learning theory	Social learning theory is the theory that people learn new behavior through observational learning of the social factors in their environment. If people observe positive, desired outcomes in the observed behavior, then they are more likely to model, imitate, and adopt the behavior themselves. Modern theory is closely associated with Julian Rotter and Albert Bandura.
Association	In quantitative research, the term "association" is often used to emphasize that a relationship being discussed is not necessarily causal .
Conflict theories	Conflict theories are perspectives in social science which emphasize the social, political or material inequality of a social group, which critique the broad socio-political system, or which otherwise detract from structural functionalism and ideological conservativism. Conflict theories draw attention to power differentials, such as class conflict, and generally contrast historically dominant ideologies.

Chapter 12. CONFLICT CRIMINOLOGY

Certain conflict theories set out to highlight the ideological aspects inherent in traditional thought.

Differential	In calculus, a differential is traditionally an infinitesimally small change in a variable. For example, if x is a variable, then a change in the value of x is often denoted Δx (or δx when this change is considered to be small). The differential dx represents such a change, but is infinitely small.
Differential association	In criminology, Differential Association is a theory developed by Edwin Sutherland proposing that through interaction with others, individuals learn the values, attitudes, techniques, and motives for criminal behavior.
	The Differential Association Theory is the most talked about of the Interactionist theory of deviance. This theory focuses on how individuals learn how to become criminals, but does not concern itself with why they become criminals.
Group	In the social sciences a group can be defined as two or more humans who interact with one another, accept expectations and obligations as members of the group, and share a common identity. By this definition, society can be viewed as a large group, though most social groups are considerably smaller.
	A true group exhibits some degree of social cohesion and is more than a simple collection or aggregate of individuals, such as people waiting at a bus stop.
Group conflict	Group conflicts is where social behavior causes groups of individuals to conflict with each other. It can also refer to a conflict within these groups. This conflict is often caused by differences in social norms, values, and religion.

Chapter 12. CONFLICT CRIMINOLOGY

Criminalization	Criminalization, in criminology, is "the process by which behaviors and individuals are transformed into crime and criminals". Previously legal acts may be transformed into crimes by legislation or judicial decision. However, there is usually a formal presumption in the rules of statutory interpretation against the retrospective application of laws and only the use of express words by the legislature may rebut this presumption.
Power	The Power of a statistical test is the probability that the test will reject a false null hypothesis (i.e. that it will not make a Type II error). As Power increases, the chances of a Type II error decrease. The probability of a Type II error is referred to as the false negative rate (β). Therefore Power is equal to $1 - \beta$.
Probability	Probability is a way of expressing knowledge or belief that an event will occur or has occurred. The concept has been given an exact mathematical meaning in probability theory, which is used extensively in such areas of study as mathematics, statistics, finance, gambling, science, Artificial intelligence/Machine learning and philosophy to draw conclusions about the likelihood of potential events and the underlying mechanics of complex systems. Interpretations The word probability does not have a consistent direct definition.
Prejudice	A prejudice is a prejudgment, an assumption made about someone or something before having adequate knowledge to be able to do so with guaranteed accuracy. The word prejudice is most commonly used to refer to a preconceived judgment toward a people or a person because of race, social class, gender, ethnicity, age, disability, political beliefs, religion, sexual orientation or other personal characteristics. It also means beliefs without knowledge of the facts and may include "any unreasonable attitude that is unusually resistant to rational influence."
Poverty	Poverty is the lack of basic human needs, such as clean water, nutrition, health care, education, clothing and shelter, because of the inability to afford them. This is also referred to as absolute poverty or destitution. Relative poverty is the condition of having fewer resources or less income than others within a society or country, or compared to worldwide averages.

119

Chapter 12. CONFLICT CRIMINOLOGY

Wealth	Wealth is the abundance of valuable resources or material possessions, or the control of such assets. The word wealth is derived from the old English wela, which is from an Indo-European word stem. An individual, community, region or country that possesses an abundance of such possessions or resources is known as wealthy.
Social control	Social control refers generally to societal and political mechanisms or processes that regulate individual and group behavior, leading to conformity and compliance to the rules of a given society, state, or social group. Many mechanisms of social control are cross-cultural, if only in the control mechanisms used to prevent the establishment of chaos or anomie. Some theorists, such as Émile Durkheim, refer to this form of control as regulation.
Crime	Crime is the breach of rules or laws for which some governing authority (via mechanisms such as legal systems) can ultimately prescribe a conviction. Individual human societies may each define crime and crimes differently. While every crime violates the law, not every violation of the law counts as a crime; for example: breaches of contract and of other civil law may rank as "offences" or as "infractions".
Self-image	A person's self-image is the mental picture, generally of a kind that is quite resistant to change, that depicts not only details that are potentially available to objective investigation by others (height, weight, hair color, gender, I.Q. score, etc)., but also items that have been learned by that person about himself or herself, either from personal experiences or by internalizing the judgments of others. A simple definition of a person's self-image is their answer to this question - "What do you believe people think about you?" A more technical term for self-image that is commonly used by social and cognitive psychologists is self-schema. Like any schema, self-schemas store information and influence the way we think and remember.
Non-Hispanic Whites	Non-Hispanic Whites, Not Hispanic or Latino are people in the United States, as defined by the Census Bureau, who are of the White race but not of Hispanic or Latino origin/ethnicity. Their origins are in Europe (the vast majority), North Africa, the Middle East, and elsewhere, and date from English, French, and other European colonization that began in the 16th century, and from the massive immigration that occurred since independence in the 18th century. They are more or less equivalent to European Americans, although not exactly; "European American" in its technical definition excludes people from North Africa and the Middle East, and includes Hispanic or Latino Whites of primarily Spanish and other European background.
Cognition	Cognition is the scientific term for "the process of thought". Usage of the term varies in different disciplines; for example in psychology and cognitive science, it usually refers to an information processing view of an individual's psychological functions. Other interpretations of the meaning of cognition link it to the development of concepts; individual minds, groups, and organizations.

Chapter 12. CONFLICT CRIMINOLOGY

Chicago School	In sociology and later criminology, the Chicago School was the first major body of works emerging during the 1920s and 1930s specialising in urban sociology, and the research into the urban environment by combining theory and ethnographic fieldwork in Chicago, now applied elsewhere. While involving scholars at several Chicago area universities, the term is often used interchangeably to refer to the University of Chicago's sociology department--one of the oldest and one of the most prestigious. Following World War II, a "Second Chicago School" arose whose members used symbolic interactionism combined with methods of field research, to create a new body of works.
Human Ecology	Human ecology is the interdisciplinary or transdisciplinary study of the relationship between humans and their natural, social, and built environments.
Race	Race refers to classifications of humans into large and relatively distinct populations or groups often based on factors such as appearance based on heritable phenotypical characteristics or geographic ancestry, but also often influenced by and correlated with traits such as culture, ethnicity and socio-economic status. As a biological term, race denotes genetically divergent human populations that can be marked by common phenotypic traits. This sense of race is often used by forensic anthropologists when analyzing skeletal remains, in biomedical research, and in race-based medicine.

Chapter 13. MARXISM AND POSTMODERN CRIMINOLOGY

Self-image	A person's self-image is the mental picture, generally of a kind that is quite resistant to change, that depicts not only details that are potentially available to objective investigation by others (height, weight, hair color, gender, I.Q. score, etc)., but also items that have been learned by that person about himself or herself, either from personal experiences or by internalizing the judgments of others. A simple definition of a person's self-image is their answer to this question - "What do you believe people think about you?" A more technical term for self-image that is commonly used by social and cognitive psychologists is self-schema. Like any schema, self-schemas store information and influence the way we think and remember.
Theory	The word theory, when used by scientists, refers to an explanation of reality that has been thoroughly tested so that most scientists agree on it. It can be changed if new information is found. Theory is different from a working hypothesis, which is a theory that hasn't been fully tested; that is, a hypothesis is an unproven theory. The word theory also distinguishes ideas from practice.
Capitalism	Capitalism is an economic system in which the means of production are privately owned and operated for a private profit; decisions regarding supply, demand, price, distribution, and investments are made by private actors in the free market; profit is distributed to owners who choose to invest in businesses, and wages are paid to workers employed by businesses and companies.
Social learning	Social learning is learning that takes place at a wider scale than individual or group learning, up to a societal scale, through social interaction between peers. It may or may not lead to a change in attitudes and behaviour. More specifically, to be considered social learning, a process must: (1) demonstrate that a change in understanding has taken place in the individuals involved; (2) demonstrate that this change goes beyond the individual and becomes situated within wider social units or communities of practice; and (3) occur through social interactions and processes between actors within a social network (Reed et al., 2010).
Social learning theory	Social learning theory is the theory that people learn new behavior through observational learning of the social factors in their environment. If people observe positive, desired outcomes in the observed behavior, then they are more likely to model, imitate, and adopt the behavior themselves. Modern theory is closely associated with Julian Rotter and Albert Bandura.

Chapter 13. MARXISM AND POSTMODERN CRIMINOLOGY

Rebellion	Rebellion is a refusal of obedience or order. It may, therefore, be seen as encompassing a range of behaviors from civil disobedience and mass nonviolent resistance, to violent and organized attempts to destroy an established authority such as a government. Those who participate in rebellions are known as "rebels".
Crime	Crime is the breach of rules or laws for which some governing authority (via mechanisms such as legal systems) can ultimately prescribe a conviction. Individual human societies may each define crime and crimes differently. While every crime violates the law, not every violation of the law counts as a crime; for example: breaches of contract and of other civil law may rank as "offences" or as "infractions".
Erikson's stages of psychosocial development	Erikson's stages of psychosocial development as articulated by Erik Erikson explain eight stages through which a healthily developing human should pass from infancy to late adulthood. In each stage the person confronts, and hopefully masters, new challenges. Each stage builds on the successful completion of earlier stages.
Deviance	In statistics, deviance is a quality of fit statistic for a model that is often used for statistical hypothesis testing. The deviance for a model M_0 is defined as $$D(y) = -2[\log\{p(y|\hat{\theta}_0)\} - \log\{p(y|\hat{\theta}_s)\}].$$
Modernization	In the social sciences, modernization refers to a model of an evolutionary transition from a 'pre-modern' or 'traditional' to a 'modern' society. The teleology of modernization is described in social evolutionism theories, existing as a template that has been generally followed by societies that have achieved modernity. While it may theoretically be possible for some societies to make the transition in entirely different ways, there have been no counterexamples provided by reliable sources.
Organized crime	Organized crime are transnational, national, or local groupings of highly centralized enterprises run by criminals for the purpose of engaging in illegal activity, most commonly for monetary profit. The Organized Crime Control Act (U.S., 1970) defines organized crime as "The unlawful activities of [...] a highly organized, disciplined association [...]". Such crime is commonly referred to as the work of the Mob in the U.S.

127

Bias	In statistics, Bias is systematic favoritism that is present in the data collection process resulting in misleading results. There are several types of statistical Bias:

· Selection Bias, where there is an error in choosing the individuals or groups to take part in a scientific study. It includes sampling Bias, in which some members of the population are more likely to be included than others. Spectrum Bias consists of evaluating the ability of a diagnostic test in a Biased group of patients, which leads to an overestimate of the sensitivity or specificity of the test. Funding Bias may lead to selection of outcomes, test samples, or test procedures that favor a study's financial sponsor.

· The Bias of an estimator is the difference between an estimator's expectation and the true value of the parameter being estimated. Omitted-variable Bias is the Bias that appears in estimates of parameters in a regression analysis when the assumed specification is incorrect, in that it omits an independent variable that should be in the model.

· In statistical hypothesis testing, a test is said to be unBiased when the probability of rejecting the null hypothesis exceeds the significance level when the alternative is true and is less than or equal to the significance level when the null hypothesis is true.

· Systematic Bias or systemic Bias are external influences that may affect the accuracy of statistical measurements.

· Data-snooping Bias comes from the misuse of data mining techniques.

Cultural bias	Cultural bias is the phenomenon of interpreting and judging phenomena by standards inherent to one's own culture. The phenomenon is sometimes considered a problem central to social and human sciences, such as economics, psychology, anthropology, and sociology. Some practitioners of the aforementioned fields have attempted to develop methods and theories to compensate for or eliminate cultural bias.
Postmodernism	Postmodernism is a movement away from the viewpoint of modernism. More specifically it is a tendency in contemporary culture characterized by the problematization of objective truth and inherent suspicion towards global cultural narrative or meta-narrative. It involves the belief that many, if not all, apparent realities are only social constructs, as they are subject to change inherent to time and place.

Chapter 13. MARXISM AND POSTMODERN CRIMINOLOGY

Power	The Power of a statistical test is the probability that the test will reject a false null hypothesis (i.e. that it will not make a Type II error). As Power increases, the chances of a Type II error decrease. The probability of a Type II error is referred to as the false negative rate (β). Therefore Power is equal to $1 - \beta$.
Homicide	Homicide refers to the act of a human killing another human. A common form of homicide, for example, would be murder. It can also describe a person who has committed such an act, though this use is rare in modern English. Homicide is not always a punishable act under the criminal law, and is different than a murder from such formal legal point of view.

Chapter 14. GENDER AND CRIME

Poverty	Poverty is the lack of basic human needs, such as clean water, nutrition, health care, education, clothing and shelter, because of the inability to afford them. This is also referred to as absolute poverty or destitution. Relative poverty is the condition of having fewer resources or less income than others within a society or country, or compared to worldwide averages.
Association	In quantitative research, the term "association" is often used to emphasize that a relationship being discussed is not necessarily causal .
Bias	In statistics, Bias is systematic favoritism that is present in the data collection process resulting in misleading results. There are several types of statistical Bias:

· Selection Bias, where there is an error in choosing the individuals or groups to take part in a scientific study. It includes sampling Bias, in which some members of the population are more likely to be included than others. Spectrum Bias consists of evaluating the ability of a diagnostic test in a Biased group of patients, which leads to an overestimate of the sensitivity or specificity of the test. Funding Bias may lead to selection of outcomes, test samples, or test procedures that favor a study's financial sponsor.

· The Bias of an estimator is the difference between an estimator's expectation and the true value of the parameter being estimated. Omitted-variable Bias is the Bias that appears in estimates of parameters in a regression analysis when the assumed specification is incorrect, in that it omits an independent variable that should be in the model.

· In statistical hypothesis testing, a test is said to be unBiased when the probability of rejecting the null hypothesis exceeds the significance level when the alternative is true and is less than or equal to the significance level when the null hypothesis is true.

· Systematic Bias or systemic Bias are external influences that may affect the accuracy of statistical measurements.

· Data-snooping Bias comes from the misuse of data mining techniques.

Chapter 14. GENDER AND CRIME

Crime	Crime is the breach of rules or laws for which some governing authority (via mechanisms such as legal systems) can ultimately prescribe a conviction. Individual human societies may each define crime and crimes differently. While every crime violates the law, not every violation of the law counts as a crime; for example: breaches of contract and of other civil law may rank as "offences" or as "infractions".
Cultural bias	Cultural bias is the phenomenon of interpreting and judging phenomena by standards inherent to one's own culture. The phenomenon is sometimes considered a problem central to social and human sciences, such as economics, psychology, anthropology, and sociology. Some practitioners of the aforementioned fields have attempted to develop methods and theories to compensate for or eliminate cultural bias.
Differential	In calculus, a differential is traditionally an infinitesimally small change in a variable. For example, if x is a variable, then a change in the value of x is often denoted Δx (or δx when this change is considered to be small). The differential dx represents such a change, but is infinitely small.
Differential association	In criminology, Differential Association is a theory developed by Edwin Sutherland proposing that through interaction with others, individuals learn the values, attitudes, techniques, and motives for criminal behavior. The Differential Association Theory is the most talked about of the Interactionist theory of deviance. This theory focuses on how individuals learn how to become criminals, but does not concern itself with why they become criminals.
Theory	The word theory, when used by scientists, refers to an explanation of reality that has been thoroughly tested so that most scientists agree on it. It can be changed if new information is found. Theory is different from a working hypothesis, which is a theory that hasn't been fully tested; that is, a hypothesis is an unproven theory. The word theory also distinguishes ideas from practice.
Patriarchy	Patriarchy is a social system in which the role of the male as the primary authority figure is central to social organization, and where fathers hold authority over women, children, and property. It implies the institutions of male rule and privilege, and is dependent on female subordination.

Chapter 14. GENDER AND CRIME

Historically, the principle of patriarchy has been central to the social, legal, political, and economic organization of Germanic, Roman, Greek, Hebrew, Indian, and Chinese cultures, and has had a deep influence on modern civilization.

Feminism	Feminism refers to movements aimed at establishing and defending equal political, economic, and social rights and equal opportunities for women. Its concepts overlap with those of women's rights. Feminism is mainly focused on women's issues, but because feminism seeks gender equality, some feminists argue that men's liberation is therefore a necessary part of feminism, and that men are also harmed by sexism and gender roles.
Non-Hispanic Whites	Non-Hispanic Whites, Not Hispanic or Latino are people in the United States, as defined by the Census Bureau, who are of the White race but not of Hispanic or Latino origin/ethnicity. Their origins are in Europe (the vast majority), North Africa, the Middle East, and elsewhere, and date from English, French, and other European colonization that began in the 16th century, and from the massive immigration that occurred since independence in the 18th century. They are more or less equivalent to European Americans, although not exactly; "European American" in its technical definition excludes people from North Africa and the Middle East, and includes Hispanic or Latino Whites of primarily Spanish and other European background.
Power	The Power of a statistical test is the probability that the test will reject a false null hypothesis (i.e. that it will not make a Type II error). As Power increases, the chances of a Type II error decrease. The probability of a Type II error is referred to as the false negative rate (β). Therefore Power is equal to $1 - \beta$.
Norm	Social norms are the behaviors and cues within a society or group. This sociological term has been defined as "the rules that a group uses for appropriate and inappropriate values, beliefs, attitudes and behaviors. These rules may be explicit or implicit.
Social learning	Social learning is learning that takes place at a wider scale than individual or group learning, up to a societal scale, through social interaction between peers. It may or may not lead to a change in attitudes and behaviour. More specifically, to be considered social learning, a process must: (1) demonstrate that a change in understanding has taken place in the individuals involved; (2) demonstrate that this change goes beyond the individual and becomes situated within wider social units or communities of practice; and (3) occur through social interactions and processes between actors within a social network (Reed et al., 2010).

Chapter 14. GENDER AND CRIME

Social learning theory	Social learning theory is the theory that people learn new behavior through observational learning of the social factors in their environment. If people observe positive, desired outcomes in the observed behavior, then they are more likely to model, imitate, and adopt the behavior themselves. Modern theory is closely associated with Julian Rotter and Albert Bandura.
Cognition	Cognition is the scientific term for "the process of thought". Usage of the term varies in different disciplines; for example in psychology and cognitive science, it usually refers to an information processing view of an individual's psychological functions. Other interpretations of the meaning of cognition link it to the development of concepts; individual minds, groups, and organizations.
Postmodernism	Postmodernism is a movement away from the viewpoint of modernism. More specifically it is a tendency in contemporary culture characterized by the problematization of objective truth and inherent suspicion towards global cultural narrative or meta-narrative. It involves the belief that many, if not all, apparent realities are only social constructs, as they are subject to change inherent to time and place.

Chapter 15. DEVELOPMENTAL THEORIES

Association	In quantitative research, the term "association" is often used to emphasize that a relationship being discussed is not necessarily causal .
Differential	In calculus, a differential is traditionally an infinitesimally small change in a variable. For example, if x is a variable, then a change in the value of x is often denoted Δx (or δx when this change is considered to be small). The differential dx represents such a change, but is infinitely small.
Differential association	In criminology, Differential Association is a theory developed by Edwin Sutherland proposing that through interaction with others, individuals learn the values, attitudes, techniques, and motives for criminal behavior.
	The Differential Association Theory is the most talked about of the Interactionist theory of deviance. This theory focuses on how individuals learn how to become criminals, but does not concern itself with why they become criminals.
Theory	The word theory, when used by scientists, refers to an explanation of reality that has been thoroughly tested so that most scientists agree on it. It can be changed if new information is found. Theory is different from a working hypothesis, which is a theory that hasn't been fully tested; that is, a hypothesis is an unproven theory.
	The word theory also distinguishes ideas from practice.
Crime	Crime is the breach of rules or laws for which some governing authority (via mechanisms such as legal systems) can ultimately prescribe a conviction. Individual human societies may each define crime and crimes differently. While every crime violates the law, not every violation of the law counts as a crime; for example: breaches of contract and of other civil law may rank as "offences" or as "infractions".
Social learning	Social learning is learning that takes place at a wider scale than individual or group learning, up to a societal scale, through social interaction between peers. It may or may not lead to a change in attitudes and behaviour. More specifically, to be considered social learning, a process must: (1) demonstrate that a change in understanding has taken place in the individuals involved; (2) demonstrate that this change goes beyond the individual and becomes situated within wider social units or communities of practice; and (3) occur through social interactions and processes between actors within a social network (Reed et al., 2010).

Chapter 15. DEVELOPMENTAL THEORIES

Social learning theory	Social learning theory is the theory that people learn new behavior through observational learning of the social factors in their environment. If people observe positive, desired outcomes in the observed behavior, then they are more likely to model, imitate, and adopt the behavior themselves. Modern theory is closely associated with Julian Rotter and Albert Bandura.
Chicago School	In sociology and later criminology, the Chicago School was the first major body of works emerging during the 1920s and 1930s specialising in urban sociology, and the research into the urban environment by combining theory and ethnographic fieldwork in Chicago, now applied elsewhere. While involving scholars at several Chicago area universities, the term is often used interchangeably to refer to the University of Chicago's sociology department--one of the oldest and one of the most prestigious. Following World War II, a "Second Chicago School" arose whose members used symbolic interactionism combined with methods of field research, to create a new body of works.
Human Ecology	Human ecology is the interdisciplinary or transdisciplinary study of the relationship between humans and their natural, social, and built environments.
Homicide	Homicide refers to the act of a human killing another human. A common form of homicide, for example, would be murder. It can also describe a person who has committed such an act, though this use is rare in modern English. Homicide is not always a punishable act under the criminal law, and is different than a murder from such formal legal point of view.
Social control	Social control refers generally to societal and political mechanisms or processes that regulate individual and group behavior, leading to conformity and compliance to the rules of a given society, state, or social group. Many mechanisms of social control are cross-cultural, if only in the control mechanisms used to prevent the establishment of chaos or anomie. Some theorists, such as Émile Durkheim, refer to this form of control as regulation.
Bias	In statistics, Bias is systematic favoritism that is present in the data collection process resulting in misleading results. There are several types of statistical Bias: · Selection Bias, where there is an error in choosing the individuals or groups to take part in a scientific study. It includes sampling Bias, in which some members of the population are more likely to be included than others. Spectrum Bias consists of evaluating the ability of a diagnostic test in a Biased group of patients, which leads to an overestimate of the sensitivity or specificity of the test. Funding Bias may lead to selection of outcomes, test samples, or test procedures that favor a study's financial sponsor.

Chapter 15. DEVELOPMENTAL THEORIES

· The Bias of an estimator is the difference between an estimator's expectation and the true value of the parameter being estimated. Omitted-variable Bias is the Bias that appears in estimates of parameters in a regression analysis when the assumed specification is incorrect, in that it omits an independent variable that should be in the model.

· In statistical hypothesis testing, a test is said to be unBiased when the probability of rejecting the null hypothesis exceeds the significance level when the alternative is true and is less than or equal to the significance level when the null hypothesis is true.

· Systematic Bias or systemic Bias are external influences that may affect the accuracy of statistical measurements.

· Data-snooping Bias comes from the misuse of data mining techniques.

Capital punishment	Capital punishment is the killing of a person by judicial process as a punishment for an offense. Crimes that can result in a death penalty are known as capital crimes or capital offences. The term capital originates from Latin capitalis, literally "regarding the head" .
Cultural bias	Cultural bias is the phenomenon of interpreting and judging phenomena by standards inherent to one's own culture. The phenomenon is sometimes considered a problem central to social and human sciences, such as economics, psychology, anthropology, and sociology. Some practitioners of the aforementioned fields have attempted to develop methods and theories to compensate for or eliminate cultural bias.
Family	In human context, a family is a group of people affiliated by consanguinity, affinity, or co-residence. In most societies it is the principal institution for the socialization of children. Extended from the human "family unit" by affinity, economy, culture, tradition, honor, and friendship are concepts of family that are metaphorical, or that grow increasingly inclusive extending to nationhood and humanism.
Labelling	Labelling is describing someone or something in a word or short phrase. For example, describing someone who has broken a law as a criminal. Labeling theory is a theory in sociology which ascribes labelling of people to control and identification of deviant behavior.

Chapter 15. DEVELOPMENTAL THEORIES

Poverty	Poverty is the lack of basic human needs, such as clean water, nutrition, health care, education, clothing and shelter, because of the inability to afford them. This is also referred to as absolute poverty or destitution. Relative poverty is the condition of having fewer resources or less income than others within a society or country, or compared to worldwide averages.
Socialization	In economic discourse, socialization refers to the process of transforming an activity into a social relationship. Socialization of production and labor is a phenomenon that takes place under capitalism due to centralization of capital and in industries where there are increasing returns to scale, eventually leading to a situation where socialization of output (or surplus value) and co-operative ownership of the means of production is necessitated. Socialization of surplus output (profit) and ownership is one aspect of transitioning from capitalism to socialism.

Chapter 16. INTEGRATED THEORIES

Social learning	Social learning is learning that takes place at a wider scale than individual or group learning, up to a societal scale, through social interaction between peers. It may or may not lead to a change in attitudes and behaviour. More specifically, to be considered social learning, a process must: (1) demonstrate that a change in understanding has taken place in the individuals involved; (2) demonstrate that this change goes beyond the individual and becomes situated within wider social units or communities of practice; and (3) occur through social interactions and processes between actors within a social network (Reed et al., 2010).
Social learning theory	Social learning theory is the theory that people learn new behavior through observational learning of the social factors in their environment. If people observe positive, desired outcomes in the observed behavior, then they are more likely to model, imitate, and adopt the behavior themselves. Modern theory is closely associated with Julian Rotter and Albert Bandura.
Chicago School	In sociology and later criminology, the Chicago School was the first major body of works emerging during the 1920s and 1930s specialising in urban sociology, and the research into the urban environment by combining theory and ethnographic fieldwork in Chicago, now applied elsewhere. While involving scholars at several Chicago area universities, the term is often used interchangeably to refer to the University of Chicago's sociology department--one of the oldest and one of the most prestigious. Following World War II, a "Second Chicago School" arose whose members used symbolic interactionism combined with methods of field research, to create a new body of works.
Human Ecology	Human ecology is the interdisciplinary or transdisciplinary study of the relationship between humans and their natural, social, and built environments.
Probability	Probability is a way of expressing knowledge or belief that an event will occur or has occurred. The concept has been given an exact mathematical meaning in probability theory, which is used extensively in such areas of study as mathematics, statistics, finance, gambling, science, Artificial intelligence/Machine learning and philosophy to draw conclusions about the likelihood of potential events and the underlying mechanics of complex systems. Interpretations The word probability does not have a consistent direct definition.

Chapter 16. INTEGRATED THEORIES

Socialization	In economic discourse, socialization refers to the process of transforming an activity into a social relationship. Socialization of production and labor is a phenomenon that takes place under capitalism due to centralization of capital and in industries where there are increasing returns to scale, eventually leading to a situation where socialization of output (or surplus value) and co-operative ownership of the means of production is necessitated. Socialization of surplus output (profit) and ownership is one aspect of transitioning from capitalism to socialism.
Labelling	Labelling is describing someone or something in a word or short phrase. For example, describing someone who has broken a law as a criminal. Labeling theory is a theory in sociology which ascribes labelling of people to control and identification of deviant behavior.
Labeling theory	Originating in sociology and criminology, labeling theory was developed by sociologist Howard S. Becker. Labeling theory holds that deviance is not inherent to an act, but instead focuses on the linguistic tendency of majorities to negatively label minorities or those seen as deviant from norms. The theory is concerned with how the self-identity and behavior of individuals may be determined or influenced by the terms used to describe or classify them, and is associated with the concept of a self-fulfilling prophecy and stereotyping.
Norm	Social norms are the behaviors and cues within a society or group. This sociological term has been defined as "the rules that a group uses for appropriate and inappropriate values, beliefs, attitudes and behaviors. These rules may be explicit or implicit.
Poverty	Poverty is the lack of basic human needs, such as clean water, nutrition, health care, education, clothing and shelter, because of the inability to afford them. This is also referred to as absolute poverty or destitution. Relative poverty is the condition of having fewer resources or less income than others within a society or country, or compared to worldwide averages.
Crime	Crime is the breach of rules or laws for which some governing authority (via mechanisms such as legal systems) can ultimately prescribe a conviction. Individual human societies may each define crime and crimes differently. While every crime violates the law, not every violation of the law counts as a crime; for example: breaches of contract and of other civil law may rank as "offences" or as "infractions".
Organized crime	Organized crime are transnational, national, or local groupings of highly centralized enterprises run by criminals for the purpose of engaging in illegal activity, most commonly for monetary profit. The Organized Crime Control Act (U.S., 1970) defines organized crime as "The unlawful activities of [...] a highly organized, disciplined association [...]". Such crime is commonly referred to as the work of the Mob in the U.S.

Chapter 16. INTEGRATED THEORIES

Risk factor	In Information security, Risk factor is a collectively name for circumstances affecting the likelihood or the impact of a security risk.
Conformity	Conformity is the process by which an individual's attitudes, beliefs, and behaviors are influenced by what is conceived to be what other people might perceive. This influence occurs in both small groups and society as a whole, and it may be the result of subtle unconscious influences, or direct and overt social pressure. Conformity also occurs by the "implied presence" of others, or when other people are not actually present.

Chapter 17. ASSESSING CRIMINOLOGY THEORIES

Association	In quantitative research, the term "association" is often used to emphasize that a relationship being discussed is not necessarily causal .
Deterrence	Deterrence is a theory from behavioral psychology about preventing or controlling actions or behavior through fear of punishment or retribution. This theory of criminology is shaping the criminal justice system of the United States and various other countries.
Differential	In calculus, a differential is traditionally an infinitesimally small change in a variable. For example, if x is a variable, then a change in the value of x is often denoted Δx (or δx when this change is considered to be small). The differential dx represents such a change, but is infinitely small.
Differential association	In criminology, Differential Association is a theory developed by Edwin Sutherland proposing that through interaction with others, individuals learn the values, attitudes, techniques, and motives for criminal behavior. The Differential Association Theory is the most talked about of the Interactionist theory of deviance. This theory focuses on how individuals learn how to become criminals, but does not concern itself with why they become criminals.
Power	The Power of a statistical test is the probability that the test will reject a false null hypothesis (i.e. that it will not make a Type II error). As Power increases, the chances of a Type II error decrease. The probability of a Type II error is referred to as the false negative rate (β). Therefore Power is equal to $1 - \beta$.
Social learning	Social learning is learning that takes place at a wider scale than individual or group learning, up to a societal scale, through social interaction between peers. It may or may not lead to a change in attitudes and behaviour. More specifically, to be considered social learning, a process must: (1) demonstrate that a change in understanding has taken place in the individuals involved; (2) demonstrate that this change goes beyond the individual and becomes situated within wider social units or communities of practice; and (3) occur through social interactions and processes between actors within a social network (Reed et al., 2010).
Social learning theory	Social learning theory is the theory that people learn new behavior through observational learning of the social factors in their environment. If people observe positive, desired outcomes in the observed behavior, then they are more likely to model, imitate, and adopt the behavior themselves. Modern theory is closely associated with Julian Rotter and Albert Bandura.

Probability	Probability is a way of expressing knowledge or belief that an event will occur or has occurred. The concept has been given an exact mathematical meaning in probability theory, which is used extensively in such areas of study as mathematics, statistics, finance, gambling, science, Artificial intelligence/Machine learning and philosophy to draw conclusions about the likelihood of potential events and the underlying mechanics of complex systems. Interpretations The word probability does not have a consistent direct definition.
Adoption	Adoption is a process whereby a person assumes the parenting for another who is not kin and, in so doing, permanently transfers all rights and responsibilities from the original parent or parents. Unlike guardianship or other systems designed for the care of the young, adoption is intended to effect a permanent change in status and as such requires societal recognition, either through legal or religious sanction. Historically some societies have enacted specific laws governing adoption whereas others have endeavored to achieve adoption through less formal means, notably via contracts that specified inheritance rights and parental responsibilities.
Cognition	Cognition is the scientific term for "the process of thought". Usage of the term varies in different disciplines; for example in psychology and cognitive science, it usually refers to an information processing view of an individual's psychological functions. Other interpretations of the meaning of cognition link it to the development of concepts; individual minds, groups, and organizations.
Chicago School	In sociology and later criminology, the Chicago School was the first major body of works emerging during the 1920s and 1930s specialising in urban sociology, and the research into the urban environment by combining theory and ethnographic fieldwork in Chicago, now applied elsewhere. While involving scholars at several Chicago area universities, the term is often used interchangeably to refer to the University of Chicago's sociology department--one of the oldest and one of the most prestigious. Following World War II, a "Second Chicago School" arose whose members used symbolic interactionism combined with methods of field research, to create a new body of works.
Human Ecology	Human ecology is the interdisciplinary or transdisciplinary study of the relationship between humans and their natural, social, and built environments.

Chapter 17. ASSESSING CRIMINOLOGY THEORIES

Crime	Crime is the breach of rules or laws for which some governing authority (via mechanisms such as legal systems) can ultimately prescribe a conviction. Individual human societies may each define crime and crimes differently. While every crime violates the law, not every violation of the law counts as a crime; for example: breaches of contract and of other civil law may rank as "offences" or as "infractions".
Poverty	Poverty is the lack of basic human needs, such as clean water, nutrition, health care, education, clothing and shelter, because of the inability to afford them. This is also referred to as absolute poverty or destitution. Relative poverty is the condition of having fewer resources or less income than others within a society or country, or compared to worldwide averages.
Ecological fallacy	An ecological fallacy is an error in the interpretation of statistical data in an ecological study, whereby inferences about the nature of specific individuals are based solely upon aggregate statistics collected for the group to which those individuals belong. This fallacy assumes that individual members of a group have the average characteristics of the group at large. This can also be referred to as the fallacy of division.
Erikson's stages of psychosocial development	Erikson's stages of psychosocial development as articulated by Erik Erikson explain eight stages through which a healthily developing human should pass from infancy to late adulthood. In each stage the person confronts, and hopefully masters, new challenges. Each stage builds on the successful completion of earlier stages.
Economic development	Economic development is the increase in the standard of living in a nation's population with sustained growth from a simple, low-income economy to a modern, high-income economy. Also, if the local quality of life could be improved, economic development would be enhanced. Its scope includes the process and policies by which a nation improves the economic, political, and social well-being of its people.
Modernization	In the social sciences, modernization refers to a model of an evolutionary transition from a 'pre-modern' or 'traditional' to a 'modern' society. The teleology of modernization is described in social evolutionism theories, existing as a template that has been generally followed by societies that have achieved modernity. While it may theoretically be possible for some societies to make the transition in entirely different ways, there have been no counterexamples provided by reliable sources.
Labelling	Labelling is describing someone or something in a word or short phrase. For example, describing someone who has broken a law as a criminal. Labeling theory is a theory in sociology which ascribes labelling of people to control and identification of deviant behavior.

Mass media	Mass media refers collectively to all media technologies, including the Internet, television, newspapers, and radio, which are used for mass communications, and to the organizations which control these technologies.
	Mass media play a significant role in shaping public perceptions on a variety of important issues, both through the information that is dispensed through them, and through the interpretations they place upon this information. The also play a large role in shaping modern culture, by selecting and portraying a particular set of beliefs, values, and traditions (an entire way of life), as reality.
Capital punishment	Capital punishment is the killing of a person by judicial process as a punishment for an offense. Crimes that can result in a death penalty are known as capital crimes or capital offences. The term capital originates from Latin capitalis, literally "regarding the head" .
Punishment	Punishment is the authoritative imposition of something negative or unpleasant on a person or animal in response to behavior deemed wrong by an individual or group. The authority may be either a group or a single person, and punishment may be carried out formally under a system of law or informally in other kinds of social settings such as within a family. Negative consequences that are not authorized or that are administered without a breach of rules are not considered to be punishment as defined here.
Unemployment	Unemployment, as defined by the International Labor Organization, occurs when people are without jobs and they have actively looked for work within the past four weeks. The unemployment rate is a measure of the prevalence of unemployment and it is calculated as a percentage by dividing the number of unemployed individuals by all individuals currently in the labor force.
Conflict theories	Conflict theories are perspectives in social science which emphasize the social, political or material inequality of a social group, which critique the broad socio-political system, or which otherwise detract from structural functionalism and ideological conservativism. Conflict theories draw attention to power differentials, such as class conflict, and generally contrast historically dominant ideologies.

Certain conflict theories set out to highlight the ideological aspects inherent in traditional thought.

Norm

Social norms are the behaviors and cues within a society or group. This sociological term has been defined as "the rules that a group uses for appropriate and inappropriate values, beliefs, attitudes and behaviors. These rules may be explicit or implicit.

Lightning Source UK Ltd.
Milton Keynes UK
UKOW01f0130250214

227084UK00001B/6/P